BAROGRAPHS

Also by the author and published by Baros Books

Aneroid Barometers and their Restoration

Bizarre Barometers and Other Unusual Weather Forecasters

Care and Restoration of Barometers, 2nd edition

FitzRoy and his Barometers

Also published by Baros Books

Antique Barometers: An Illustrated Survey, 2nd edition by Edwin Banfield

The Banfield Family Collection of Barometers by Edwin Banfield

Barometer Makers and Retailers 1660–1900 by Edwin Banfield

Barometers: Aneroid and Barographs by Edwin Banfield

Barometers: Stick or Cistern Tube by Edwin Banfield

Barometers: Wheel or Banjo by Edwin Banfield

The History of the Barometer by W. E. Knowles Middleton

The Italian Influence on English Barometers from 1780 by Edwin Banfield

A Treatise on Meteorological Instruments by Negretti & Zambra

BAROGRAPHS

Second edition

Philip R. Collins

Baros Books

© Philip R. Collins 2002, 2021
First edition published 2002
Second edition published 2021

Baros Books
27 Queen Square
North Curry
Taunton
Somerset
TA3 6LE
UK
www.barosbooks.co.uk

Philip R. Collins has asserted his right under the Copyright, Designs and Patents Act, 1988 to be identified as Author of this work.

All rights reserved. No part of this publication may be reproduced, stored in a retrieval system, or transmitted, in any form or by any means, electronic, mechanical, photocopying, recording or otherwise, without the prior permission of the publisher.

A CIP catalogue record for this book is available from the British Library
ISBN 978-0-948382-17-8

Typeset in 12/13 pt Garamond by George Ashton (geo.ashton@gmail.com)

Printed and bound in Great Britain by Short Run Press Ltd, Exeter, Devon

Cover image: Oak-cased dial barograph by Short & Mason, c.1910.

Cover design: George Ashton (geo.ashton@gmail.com)

To 'Grandad' with grateful thanks for many hours in the dark room

Contents

Foreword		viii
Preface		ix
Acknowledgements		x
1	The First Barographs and Self-recording Aneroids	1
2	Aneroid Barographs pre-1902	26
3	Aneroid Barographs post-1902	55
4	Unusual Barographs and Altitude Recorders	98
5	General Maintenance and Repair	133
6	Bellows, Nibs, Charts and Ink	145
Appendix: Summary of Patents		167
Bibliography		169
Index		171

Foreword to the First Edition

The barograph, like the barometer, is a scientific instrument which we have allowed to enter the home as a decorative, as well as useful, piece of engineering, assisting us to continue our obsession with the weather. The purpose of both instruments is to indicate the changing pressure of the atmosphere, thus helping us to forecast probable changes in the weather. But the barograph has an advantage over the barometer: it records a continuous graph of air-pressure changes which makes a fascinating visual record and never fails to surprise us with its peaks and troughs.

In this book, Philip Collins traces the development of the barograph from the earliest self-recording aneroids, through their heyday and into post-war times, in his usual dissecting manner, using numerous examples as well as information from old catalogues and patents. His thorough discussion will help the reader in the dating of barographs which has often been difficult up to now. This book should therefore become a useful work of reference for collectors and dealers alike.

Edwin Banfield

Preface

From the beginning of my research for the first edition of this book, it was my aim to portray that most fascinating of instruments – the barograph – using illustrations as examples of the many different types still available. It became almost a detective hunt to trace their development, by inspecting existing instruments and consulting catalogues and patents, a hunt which included the famous Alexander Cumming barograph at Buckingham Palace and its little-known sister on the Isle of Bute in Scotland. I learnt far more about barographs than I had dreamed possible (and I even dreamed about them while writing the book!), while realising at the same time how little we actually know about the people involved in their development. Perhaps the most surprising find was to discover that the French were probably the greatest makers of this seemingly most British of instruments during its development in the late Victorian era.

When new, barographs, because of the nature of their construction, were always more expensive than aneroid barometers. They have reduced significantly in price over recent years, but are still sought after as highly desirable display objects for people who appreciate old instruments. Despite being generally more expensive to collect than aneroid barometers, there is a sufficient number still available to offer the would-be collector or one-off purchaser a wide choice of styles and prices. I hope this second edition will help the reader to identify many of the barographs still to be found and encourage more people to acquire a good one to use and enjoy.

Inevitably, there will be some gaps in coverage as the content of this book is chiefly founded on examples handled over a number of years in my own workshop, as well as on instruments formerly in the Banfield Family Collection of Barometers and others belonging to private collectors, dealers, friends and associates.

Acknowledgements

My grateful thanks are due again to the late Edwin Banfield for his help and advice in publishing the first edition of this book. His books and former barometer collection played a large part in kindling my own passion for barometers and barographs. I would like to thank his daughter Sue Ashton and his grandson George Ashton for their work on this second edition.

Thanks are also due to the Patent Office staff and past and present members of the National Meteorological Office Library and Archive for the use of their records, and to the many collectors, colleagues, curators and dealers, as well as surviving companies, who have all helped me to produce this book.

Last, but by no means least, I thank my father for his many hours in the dark room producing nearly all of the photographs for the first edition of the book, a task now superseded by a digital camera so he can have a much-deserved rest.

Illustrations

The author and publisher would like to thank Richard Twort for his permission to use illustrations of items throughout the book. They also gratefully acknowledge the permission of the following to reproduce illustrations:

Fig. 1.1: Mr P. Dixon
Figs 1.3, 1.4 and 1.5: Mr J. F. M. de Vree
Figs 1.6 and 2.19: Mr P. D. Bosson
Figs 1.7 and 4.20: Casella
Fig. 1.9: © Crown Copyright 1868. Information provided by the National Meteorological Library and Archive – Met Office, UK
Figs 1.10 and 1.11: © Royal Meteorological Society
Fig. 1.14: Mr P. Negretti
Fig. 1.16: © National Museums Scotland
Fig. 1.18: Dreweatts 1759

Fig. 2.2: Baskerville Antiques
Figs 2.17 and 2.18: Davis of Derby Limited
Figs 4.18 and 4.19: Negretti Automation
Figs 4.40 and 4.41: Mr J. Whitworth

Every effort has been made to trace copyright holders, but if any has been inadvertently overlooked, the publisher will be pleased to make the necessary arrangement at the first opportunity.

1 The First Barographs and Self-recording Aneroids

A barograph is a barometer that is able to record a history of past air-pressure changes. Barographs are mostly associated with a line drawn on graph paper, but can use other methods to achieve a recording, and many different methods are shown in the following pages. They were also sometimes referred to as 'self-recording' barometers or aneroids. In W. E. Knowles Middleton's book, *The History of the Barometer* (1968), there are numerous references to barographs of many weird and wonderful types, but disappointingly – though understandably – his book has almost no illustrations of antique collectable barographs such as the type you will see illustrated within the pages of this book. He does describe, however, through line drawings and text, many different types, but few of these I intend to cover here.

There is little doubt that the first recording instrument that can be called a barograph was the 'weather clock' or 'weatherwiser' begun by Christopher Wren and improved and added to by Robert Hooke before 1681. Amongst Hooke's additions was a barometer that served to show the difference between the greatest and the least pressure of the year. Regrettably, there is no known trace of this early instrument, though it can be assumed that it worked from a mercury siphon tube.

Mercury Barographs

The oldest surviving barograph (see *Fig*. 1.1) was made for George III in 1765 by Alexander Cumming (1733–1814), a famous Scottish clock and instrument maker who moved to England and had premises in Bond Street. A staggering £1,178 was paid for this barograph. It is housed at Buckingham Palace and is undoubtedly the most superior barograph anywhere in the world. I do not doubt that there may be more accurate ones, but this example outweighs any competition as a domestic barograph. This most exquisite clock/barograph is veneered in kingwood with extensive ormolu mounts and stands 7 feet 11¾ inches (2.4 m) tall. It is surmounted by two ormolu figures of unknown significance, the larger pointing downwards, his index finger curled, and the smaller (child) looking towards the adult. Between them is a globe decorated with small punch marks for the outlines of the continents and around the centre of the equator is a barograph-type chart or line.

The mechanism of this instrument is quite fascinating and of exceptional quality. It operates from two glass tubes with flattened spherical reservoirs at the top, each of which is a vacuum. These are positioned inside two carved ivory pillars and are connected at the base to an ivory well. These twin siphon tubes communicate into the well, which measures over 2 inches (5 cm) in diameter. Plenty of mercury therefore supports the cradle arrangement, which transfers the movement of the mercury to the pencil.

This cradle is constructed in hexagonal form utilising ebony rods kept apart by ivory frames, one at the base, three visible through the glass front door and one just behind the barometer dial. The base of this cradle device sits within the ivory well and has six ivory wheels attached to it to guide along the sides as it runs up and down. The barometer is operated by a simple pulley wheel, which has a fine thread wrapped around it at one end tied to a counter-balance that hangs freely, the other end being attached to the cradle. As this cradle goes up and down, it turns the wheel, which is directly behind the centre of the hand. The barometer dial is enamel and divided from 28 to 31 into inches and hundredths

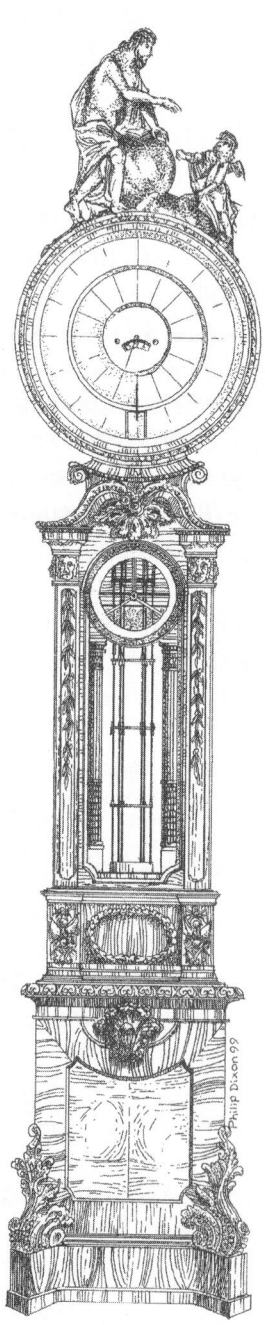

Fig. 1.1 Barograph made by Alexander Cumming for George III, finished in 1765.

of inches of mercury. The cradle extends upwards towards the vellum charts, one that rotates annually and the other half-annually through the gearing of the clock.

The charts are possibly original, being considerably faded now, but quite clearly hand-divided and hand-written in black ink with black, blue and red divisions intermixed. The metal extension from the top of the front two ebony rods that emerge from the cradle consists of two pieces of metal 1½ inches (3.8 cm) apart, to which is screwed a turned metal piece which makes a frame. This cross-piece has a central screw to secure a blued steel indicator, which is of fine turned quality, having points top and bottom. This is adjustable with the screw so that it can be positioned higher or lower. It is a replacement of an earlier pencil-mounted device. It may be that these charts were re-useable as there are three charts surviving which were transcribed from this barograph in 1765, 1766 and 1767, and which have been drawn in the more linear and rectangular fashion that we are used to in modern barograph charts. Graphite as used in pencils has been available since the very early sixteenth century. The charts are held on to their metal backing frame, which has gears to rotate them, by outer metal rims which are pinned into position on the reverse.

The diameter of the mercury column being about 2 inches (5 cm), the reservoir would certainly lift the cradle, which weighs somewhere in the region of 1¼–1½ lbs (570–680 g), up and down quite easily. In the manuscript notes by Alexander Cumming for a new edition of his book *Elements of Clock and Watchwork*, held in the Guildhall Library, he says of the chart that 'this journal may be contained on the same piece of vellum for some years': so we see that it was intended that the charts should last some time. The Science Museum has a chart which has three years recorded whilst owned by Luke Howard in 1818, 1842 and 1843, so they were obviously re-used. Interestingly, the charts in the Science Museum appear to be inked, leading one to assume that they were inked in after being used. I suspect, as with most early barographs, that considerable difficulty was found in recording the pressure on the charts due to friction. The very large diameter of mercury and therefore the ease of movement of the cradle up and down would probably overcome most of this.

The two ivory pillars, which rise up either side and just behind the cradle, are united at the top by a piece of ivory some 4 inches (10 cm) deep, intricately carved although little seen behind the barometer dial. Continuing above this, in line with the ivory columns, can be seen, if looked at very carefully at an angle with a torch, the top of the glass columns, which are a flattened, rounded shape approximately 3 inches (7.5 cm) in diameter. The ormolu mounts were possibly re-gilded in the early nineteenth century, but the whole item is in very fine condition. It was exhibited at the Science Museum, London in 1952 and at The Queen's Gallery, Buckingham Palace in

1974–5. On the occasions that Buckingham Palace is open to visitors, the tour normally includes the area where this barograph stands.

A second barograph by Alexander Cumming is similar, although less ornate, and can be viewed at the Science Museum, London. Cumming made it in 1766 for his own use, but his relatives (after his death in 1814) allowed Luke Howard to purchase the barograph and he seems to have made good use of it. Howard (1772–1864), a chemist with a pharmacy in Fleet Street, was a keen amateur meteorologist and is probably most famous for giving names to clouds which are still in use today. He commented in 1818 that Cumming's barograph required no more than winding of the clock, and he published data about the pressures he recorded in London. He also described the barograph as recording by use of a pencil (quoted in *Weather*, August 1952, p. 252).

This model had only one yearly barograph dial, presumably again of vellum. The barograph stands 7 feet 1 inch high (2.16 m). The outside diameter (width) of the dial and case is about 20 inches (50 cm), the widest part at its base about 24 inches (60 cm). Made of mahogany, it is decorated with applied carvings and veneer. The inner clock dial, inscribed 'Alex R. Cumming, London 1766', has two winding holes. The central dial shows minutes; the second outer dial is divided into 24 hours; the third dial holds the barograph chart similar to the royal barograph. This chart is a modern replacement dated 1975: it is divided from January to December with a small pencil-holder held on top of a floating cage by a mahogany stick and brass mount. The glazed door is similar to a long-case clock, revealing a pendulum with a mirror at the back, and two carved mahogany Corinthian columns, with an ivory and wood floating cage.

The design of this instrument differs from the royal barograph somewhat. The ivory well has a cover to it and in the centre rises one piece of mahogany, the lower end obviously intended to rest on mercury. Around the outside of this cover are six ivory pegs, which support three ivory wheels that guide the mahogany rod up and down. On top of the first mahogany rod is a horizontal piece of ivory like a candelabra; on its left and right rise two mahogany rods and in the centre is an ivory thermometer scale about 9 inches (23 cm) tall. These two mahogany rods are held at their tops by another similar horizontal piece of ivory, which then supports a single central mahogany rod, which rises up towards the chart and on top of which the recording device is held. The two rods side by side are thinner than the single rods, each about ¼ inch (6 mm) in diameter. The two Corinthian pillars hold glass tubes in a similar fashion to the royal barograph. The carved panel at the front at the base of the barograph is hinged, obviously to gain access. For those interested in the clock mechanism, it is actually fitted with a gravity escapement which, it is suggested, was the first ever made and invented by Alexander Cumming, as described and illustrated in his book

Elements of Clock and Watchwork published in 1766.

A third extant barograph by Alexander Cumming can be found in Scotland at Mount Stuart House on the Isle of Bute. This incredible house is open to the public all year round. The barograph is certainly worth a visit, and if you are able to travel there I recommend it. For readers who are also interested in clocks, there are a number of exceptionally fine clocks on display as well. The barograph is in an elaborately carved and decorated mahogany case with one 12-month dial to record the pressure, housed around a clock dial with seconds, minutes and hours. The connection to the graph dial is by a small gear driven from the clock to a rack fixed to the back of the barograph dial.

The barometer dial, from 28 to 31 inches, inscribed 'Alex R. Cumming, London' in black letters on a white enamel background, is almost identical to the royal barograph. It has two glass barometer tubes, which appear to be similar in size to the royal one, but housed in mahogany pillars not ivory. The ivory well has an internal bore of 2½ inches (6 cm) and retains its original top. There is a mahogany rod going through the centre of the well cover and connecting to the bottom of the cage, with an ivory plunger piston to rest on the mercury. It is fitted with a hexagonal cage similar to the royal barograph, which terminates in a slightly different design – two pillars extending up, with a third one at the front about an inch (2.5 cm) long and hollow, which holds the writing device. There appears to be a simple metal wire which acts as a spring and runs over a very small (⅛ inch/3 mm) pulley wheel, which is designed to keep the pencil in contact with the chart with enough pressure to record but not restrict the operation of the instrument. The metal circular barograph chart holder is 18 inches (45 cm) in diameter. Its inner ring is inscribed with the abbreviated months of the year and the number of days in the months in between each of these. The total height of the barograph is 8 feet 9 inches (2.67 m). It is very similar in mechanical design to the royal clock/barograph, but its mahogany case is much more elaborate than the one at the Science Museum.

A fourth barograph by Alexander Cumming is held at the Victoria and Albert Museum, London. It is similar in design to the one at Mount Stuart House on the Isle of Bute in Scotland but, sadly, all of its barometer parts are missing except for the enamel dial. According to records held by the V&A, it was made around 1774 for Sir James Lowther. (He had married Lord Bute's daughter in 1761 and would have known of the barograph owned by Lord Bute.) A fifth known barograph by Alexander Cumming differs from the others in being a wall-mounted clock and barometer measuring 4 feet 4 inches high by 16¾ inches wide (1.3 m x 43 cm). It has the same principal design of two mercury tubes combining into a common reservoir to lift a rod so as to indicate pressure changes on a semi-circular dial and annual chart. It was sold by Sotheby's New York on 25 January 1997.

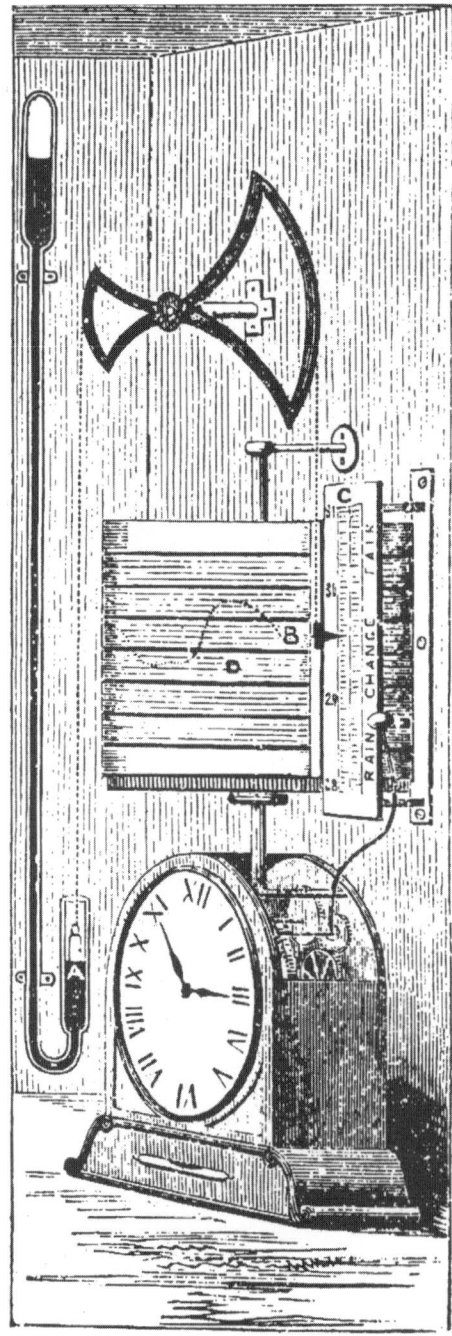

Fig. 1.2 Milne-style barograph as illustrated by Negretti & Zambra in 1864.

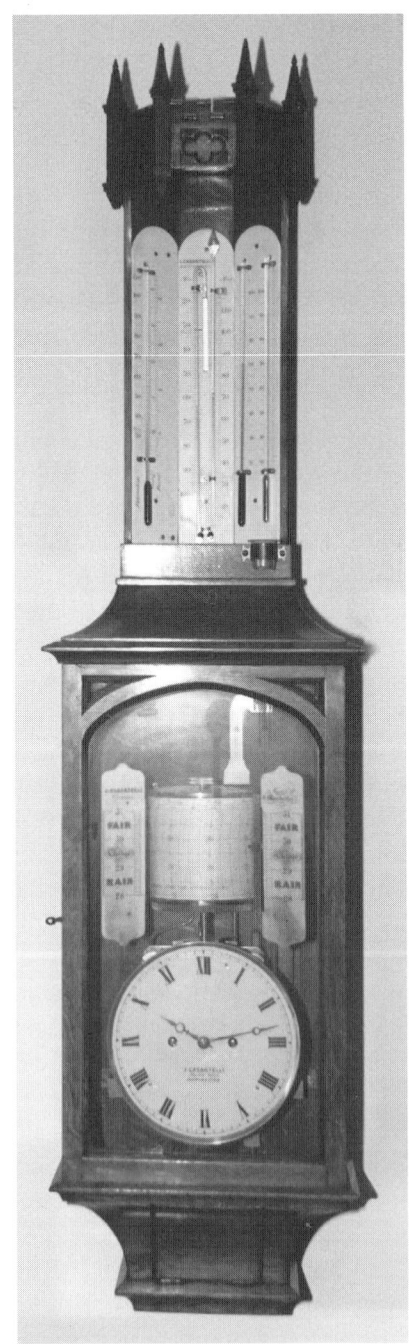

Fig. 1.3 Oak-cased siphon tube barograph by J. Casartelli, c.1865.

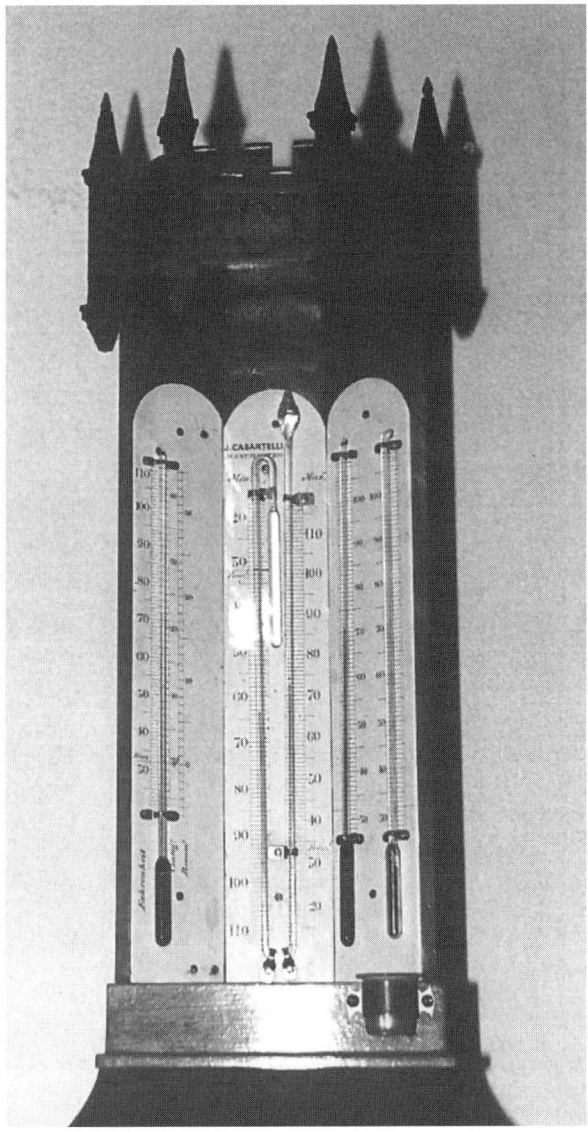

Fig. 1.4 Thermometers on oak barograph by J. Casartelli (*Fig.* 1.3).

A number of other barographs were designed in Europe, all of which had to be operated from a mercury tube until the aneroid mechanism was invented by Lucien Vidi in 1844. Of the mercury-type barograph, very few survive and they are exceptionally difficult to purchase. A type that does occasionally come up on the market is similar in design to the Admiral Sir Alexander Milne (1806–1896) style barograph, as described and made by Negretti & Zambra (see *Fig.* 1.2). The design of these, however, bears little

Fig. 1.5 Clock and recording drum on oak barograph by J. Casartelli (*Fig.* 1.3).

actual resemblance to the barograph made by Admiral Sir Alexander Milne other than the use of a siphon tube. They utilise a fairly standard type of siphon tube and a counter-balanced lever and a clock to drive the recording drum around. As the float in the tube rises up and down, the other end of the balance lever moves the pencil up and down in front of the paper chart. By an arrangement with the clock, the pencil makes a mark on the chart every half-hour.

The First Barographs

Fig. 1.6 Oak siphon tube barograph by J. Casartelli with single thermometer, c.1870.

The ones I have seen of this type are generally of oak. They have varying cases but are similar to the one illustrated in *Fig.* 1.3 by J. Casartelli of 43 Market Street, Manchester (working dates 1852–1896). It probably dates from the mid-1860s. The case has a very gothic look with rising pinnacles above four thermometers (shown in *Fig.* 1.4), including Fahrenheit, centigrade and Reaumur scales, alongside a minimum/maximum thermometer and a wet dry bulb hygrometer. In *Fig.* 1.5 the pencil arrangement can be seen to the right of the recording drum, which is very similar to the self-recording aneroid-style barograph discussed below. To the left and right are two engraved ivory scales which line up with the chart and have the same 'weather words' engraved on them as nearly all stick barometers of the time. The whole barograph measures just over 48 inches (1.2 m) high and is in very good condition. *Fig.* 1.6 shows another mercury recording barograph by Casartelli. All these types are rare, but this instrument depicts the style that is most often found. It is again oak-cased with a hinged glazed door to allow winding of the clock and changing of the charts.

The instrument in *Fig.* 1.7 is a notable self-registering barometer by engineer Alfred King of Liverpool. It was designed in 1854 to register by continuous pencil tracing the variations in the weight of the atmosphere. The whole tube of mercury moves up and down, and for every 1 inch (2.5 cm) movement of mercury the pencil moves through 5 inches (12 cm). There is an excellent example of this instrument in the Science Museum in London (not currently on display). The illustration comes from Casella's catalogue number 498, which is after 1911, probably just

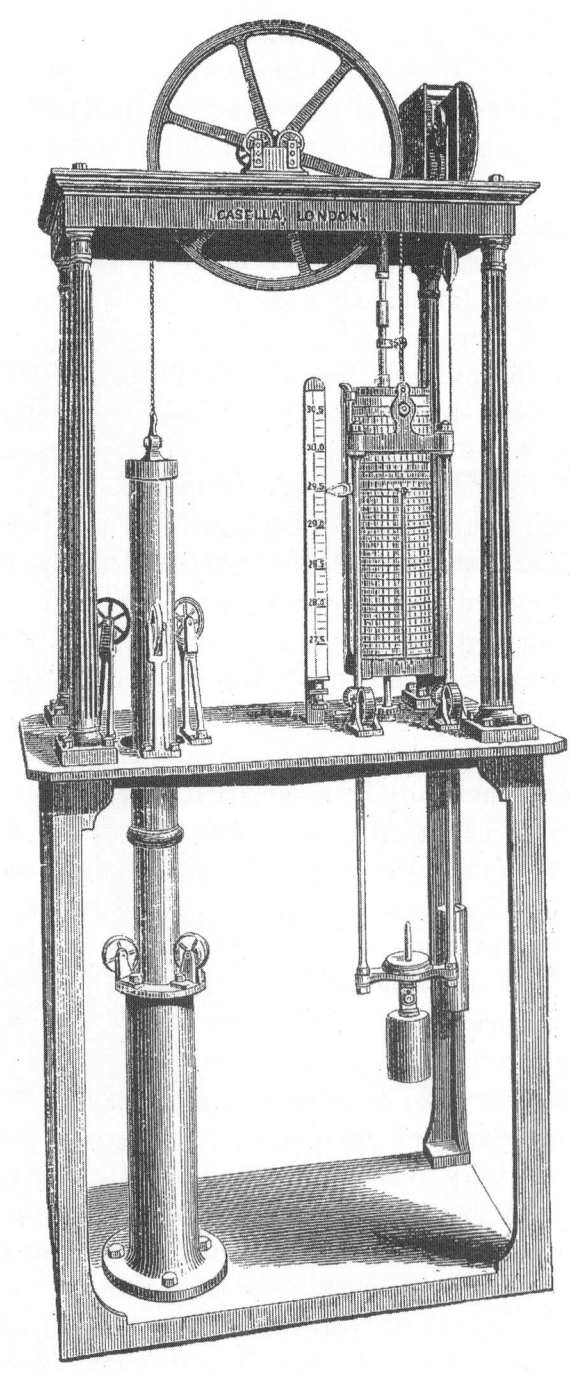

Fig. 1.7 An Alfred King barograph, designed in 1854, as illustrated in a Casella catalogue, c.1914.

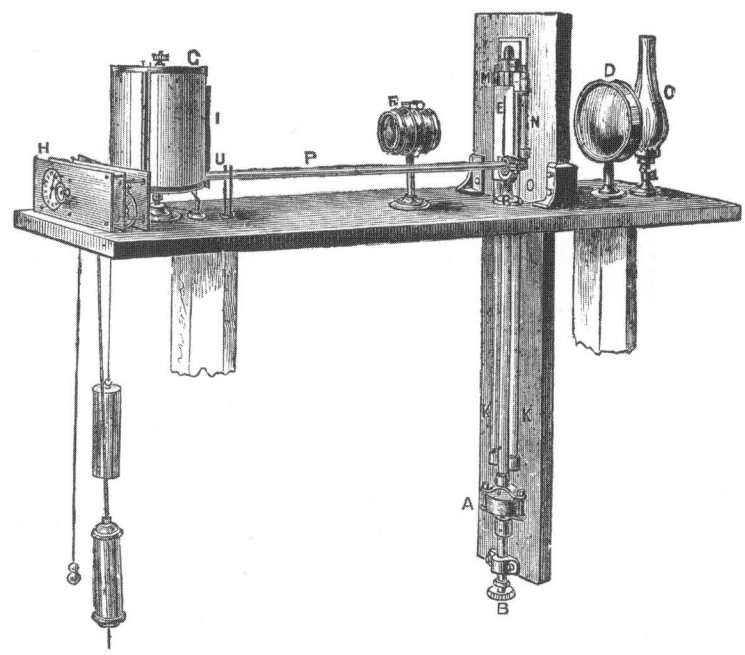

Fig. 1.8 Kew station barograph designed around 1851.

before or after the First World War. It is interesting to note that they were still advertised for sale then, as was the Kew barograph, but I imagine that they were soon phased out in favour of other types.

In 1839, T. B. Jordan described the first photographic recording of barometric pressure in the Royal Cornwall Polytechnic Society's Annual Report. Based on photographic paper prepared according to the directions of H. Fox Talbot (1800–1877), it utilised a stick barometer tube housed in a wooden case with photographic paper revolving around the clock, which allowed light to shine over the level of mercury and record on the photographic paper. Originally it was designed to be used during daylight. Later in 1839, Professor J. P. Nichol referred to Jordan's photographic work, and suggested that meteorological instruments might utilise the same principle.

By 1851, the Kew barograph had been designed and, with a number of alterations for temperature compensation, it became the standard instrument for recording pressure. There are very few of these surviving. An example of this extremely well-engineered instrument can be seen in *Fig.* 1.8. As a modification to the original design, zinc rods were used to turn a long rod along the top of the bed and to record temperature on the bottom of the graph. This was, in effect, a temperature and pressure recording instrument, the temperature being used as a correction factor. The clockwork

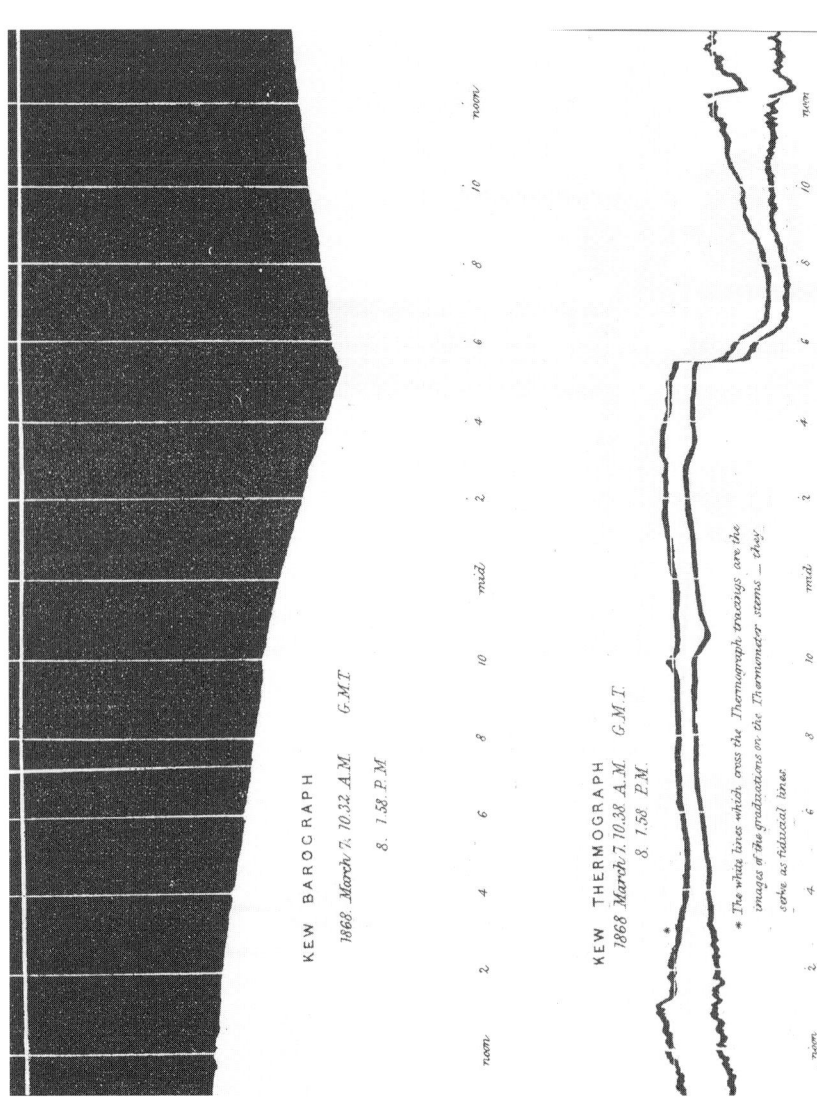

Fig. 1.9 Copy of a developed chart from a Kew station barograph dated 1868. © Crown Copyright 1868. Information provided by the National Meteorological Library and Archive – Met Office, UK.

mechanism, being a two-day movement, turned the drum holding photographic paper.

A light source, originally an oil lamp, shone light through the condensing lens across the mercury tube and was focused with a focusing lens onto the photographic paper. Consequently, the resulting chart, once developed, showed black where the vacuum was and white where no light penetrated. A copy of a chart produced by this method on 7 March 1868 is shown in *Fig.* 1.9. Naturally, the whole instrument was enclosed in a dark cabinet, with a door to allow changing of the photographic paper. Every two hours the clock moved a lever in front of the light, which gave a two-hour time check on the chart. This lasted for two minutes to create a vertical white line on the chart and it was during this two-minute period that a replacement chart had to be fitted. For ease, these were pre-loaded onto a secondary drum, the drums were changed and the clock was wound by pulling a wire to raise the weight.

Another mercury barograph of particular note was the William Henry Dines recording barometer of 1904. We are fortunate to be able to refer to a detailed description of this particular instrument by his son L. H. G. Dines in which we can begin to see some of the problems encountered by designers at this time:

> On more than one occasion attempts have been made to get the Dines float barograph made by commercial instrument makers. Unfortunately, they seem frequently to find considerable difficulty in turning out a satisfactory instrument if it were of an unorthodox pattern, even though no recondite principles are involved in its construction. It is a pity that such should be the case, and it is to be hoped that in future it will not be necessary to make such a charge against British instrument makers. (Quoted in *Journal of the Royal Meteorological Society*, vol. 55, no. 229, January 1929)

The description details not only the design but also the calculation for various compensations.

The difficulty in getting glass tubes of accurate and equal diameter was also a problem. *Fig.* 1.10 is a diagram of the working parts. It consists chiefly of two siphon tubes joined at the top end so that mercury can flow through both; this enables accurate filling in order to obtain a good vacuum when installed and is easily refilled whenever required. The accuracy of the two large reservoirs 'S' needed to be very good. A float resting on the mercury was made from a cup of glass, which trapped air inside it. This was a fundamental invention because it created a good and simple method of temperature correction. Ball bearings were also floated around this cup to

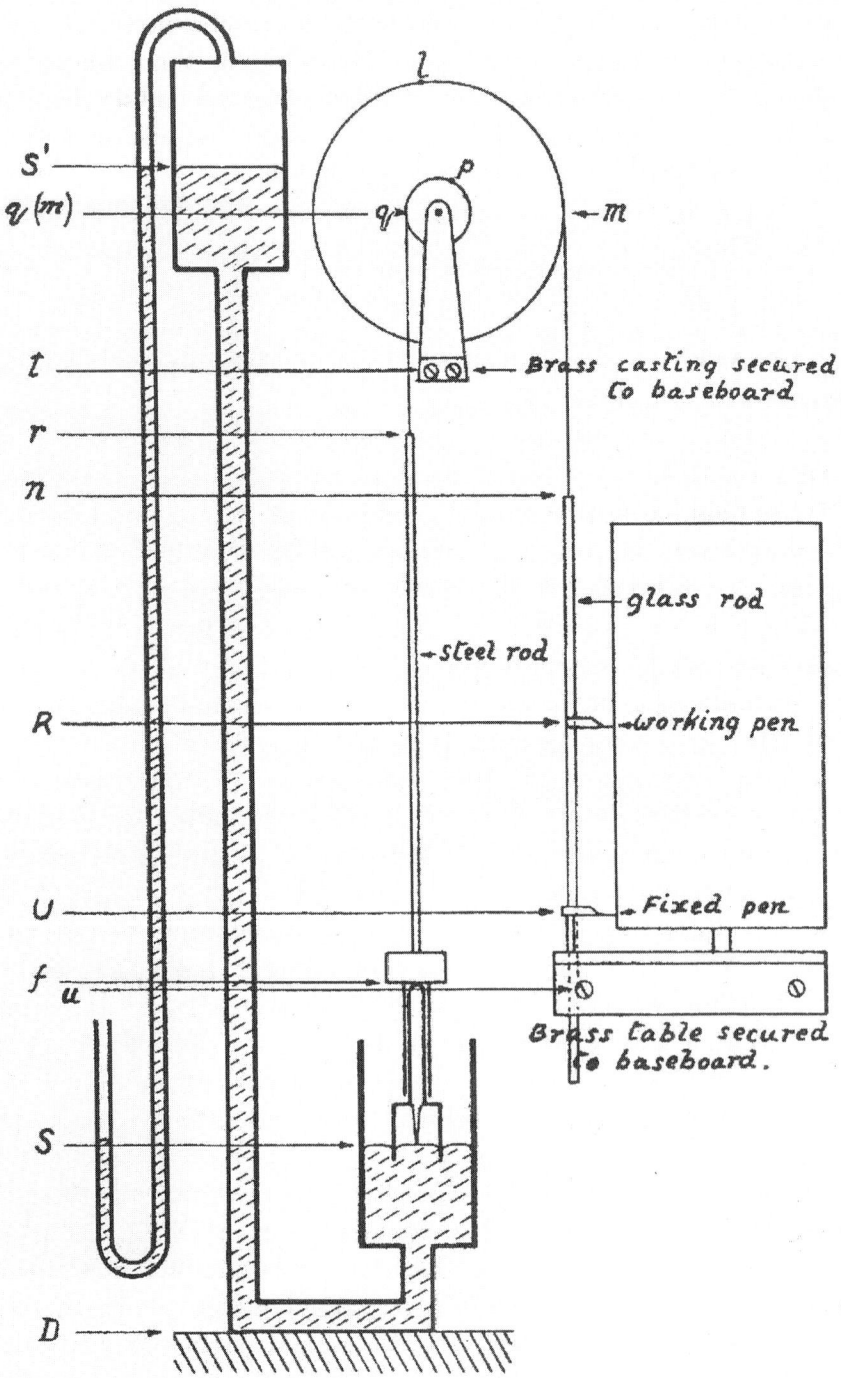

Fig. 1.10 Diagram of the Dines float barograph of 1904.
© Royal Meteorological Society.

hold it centrally within the mercury reservoir. Glass rods (which have little if any thermal expansion) connected the float to platinum wires, which went to the concentric pulley wheels, which lifted or dropped the pen. The ratio of the wheels allowed sufficient magnification (2 to 1) of the movement of mercury to make a very sensitive trace of the pressure. Its main features were considered to be a very stiff control of the pen, an almost entire absence of friction in the moving parts and an exceedingly small error due to capillary action.

A mercury tube was also utilised to record pressure in the Atmospheric Recorder, illustrated in *Fig.* 1.11, attributed to G. Dollond (1774–1856) and shown at the Great Exhibition of 1851 (see Insley 2000, p. 254). The frame which supports the chart is 2 feet wide and 3 feet 6 inches long (60 cm x 1 m) and stands on four legs. It is a complicated weather recorder and I doubt if it survived. The barograph is operated by the siphon tube in the foreground.

Self-recording Aneroid Barometers

After the self-recording mercury barometer came the first aneroid-type barographs, which were more generally known as 'self-recording aneroid barometers'. It is a name that suitably covers those early aneroid barographs before the word 'barograph' became the generally accepted term. *Fig.* 1.12 shows the earliest known self-recording aneroid barometer, made by the Parisian firm Breguet and exhibited at the Paris International Exhibition of 1867 (see Middleton 1968, p. 425). It is recorded that the Breguet business of manufacturing clocks and watches was sold outright to Edward Brown (1829–1895) in 1867 (see Brenni 1996a, p. 21), the same date that Middleton suggests that this barograph (*Fig.* 1.12) was made. We do not know, therefore, whether this model was successful or not, although they are not encountered often. It is likely that the smoked paper was a problem, especially to any potential domestic market. It has a clock which is separate from the chart drum, which also displays the time, and the charts are marked by a point inscribing a mark on smoked paper. There may well have been other makers of this type of instrument but I have not come across any. Documents are scarce and the usual early instruments found are the self-recording ones advertised by various firms from around 1870 onwards.

Fig. 1.13 shows an ebonised self-recording barometer, c.1875, by Mottershead & Co. of Manchester. Similar to many of the period, this one measures 26½ inches by 8 inches by 16 inches high (67 x 20 x 40 cm). The glazed front hinges forward and down and is locked by a key. The clock on the left drives the drum in the centre, which holds the paper chart. The aneroid movement on the right extends a secondary fusee chain, which is connected to a lightweight carrier of a pencil lead, which freely slides up

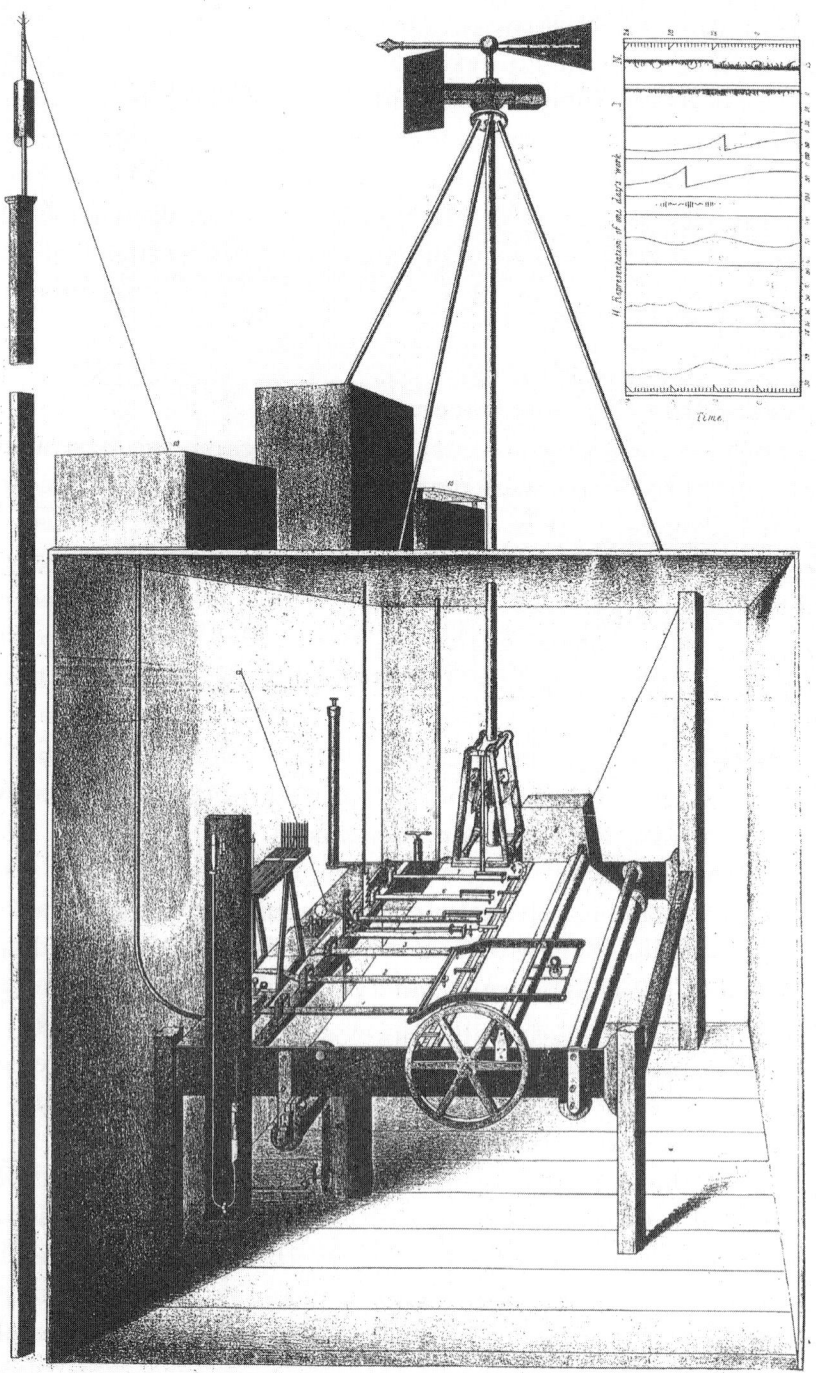

Fig. 1.11 The Atmospheric Recorder of G. Dollond, 1851.
© Royal Meteorological Society.

The First Barographs

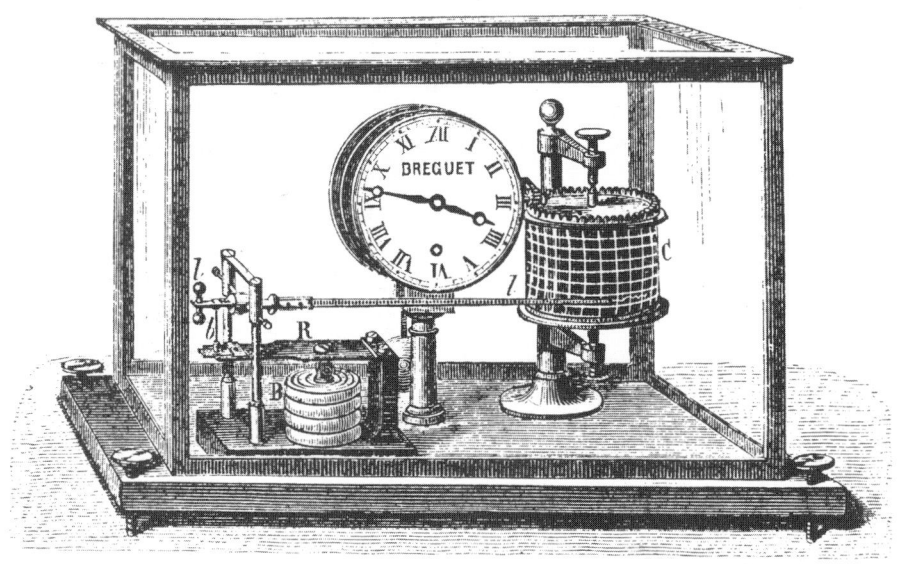

Fig. 1.12 Earliest known illustration of the common type of barograph by Breguet of Paris, c.1867.

Fig. 1.13 Ebonised self-recording barometer by Mottershead & Co. of Manchester, c.1875.

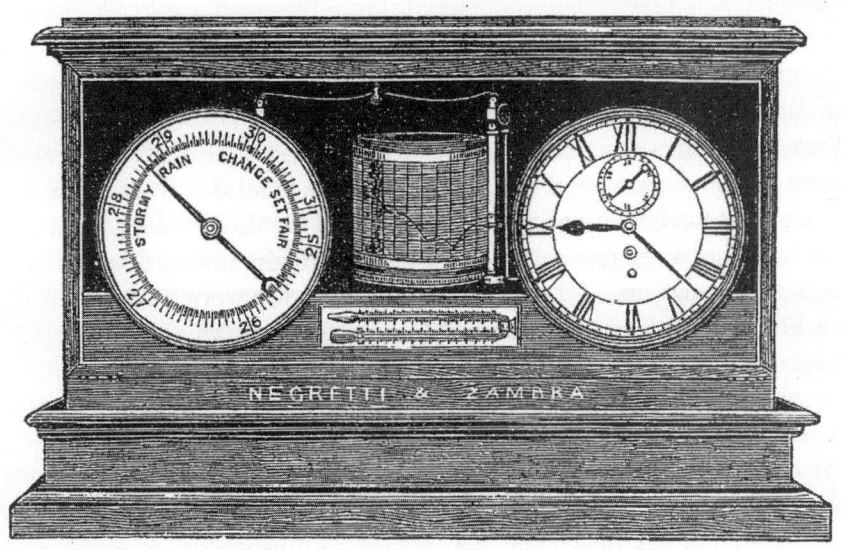

Fig. 1.14 Self-recording barometer illustrated by Negretti & Zambra in their 1873 catalogue.

and down a brass column when the pressure rises or falls. This pencil is continually rising or falling free of the chart. It also has a silvered engraved brass scale to the right of the pencil.

Five minutes before each half-hour, the clock operates a hammer to tap the whole metal plate that the drum sits on. This vibrates the aneroid movement to take out some of the slack, and at the half and full hours the clock operates a cam to rotate the pencil and make contact with the chart, after which it moves away from the chart. Below the centre drum should be mounted a minimum/maximum thermometer. This one is missing. The mechanism on this type of instrument is invariably of exceptionally high quality and would have been quite expensive. Negretti & Zambra were advertising this style for sale in their 1873 catalogue (see *Fig.* 1.14), priced at £22, and also only this type in their 1879 catalogue. The description from the catalogue reads as follows:

> This instrument is arranged to show the various fluctuations that have taken place in the barometer during the absence of the observer. It consists of a large and carefully finished aneroid and an eight-day clock; between these is placed in a vertical position a revolving cylinder, having a metallic paper attached to it, ruled to coincide with the inches and 1/10th of the barometer scale. Close to this paper is a pencil mounted on a me-

tallic rod and is moved up and down as a variation of atmospheric pressure acts upon the vacuum chamber of the aneroid; at every hour this pencil is made to mark the paper by simple mechanism via connection with the clock. By this means a black dotted curved line is produced on the paper, showing at a glance the present height of the barometer – whether it is falling or rising – for how long it has been doing so, and at what rate the change has taken place – if falling or rising at the rate of 1/10th of an inch per hour, or 1/10th in 24 hours; all of which are particularly most essential to knowing when foretelling the weather, and which can only be obtained from an ordinary barometer by very frequent and regular observations. Our engraving [*Fig.* 1.14] shows the ordinary mounting of the registering aneroid, combining a reliable time piece with an exceedingly interesting meteorological instrument of a suitable and convenient size for a library or dining room mantle shelf. £22.

The lack of surviving catalogues from many of the instrument firms in business during this period makes complete comparison impossible, and it is to be hoped that more evidence will turn up. It is likely that many firms were beginning to sell these self-recording instruments, but it is difficult to prove just which firms made them and who actually sold them. It seems likely that they were only sold for a short period of time. There is evidence that they were available in 1868 and perhaps earlier.

Patent 2924 of 1868 mentions aneroids for use in recording pressure but describes Bourdon tube recorders mostly. Bourdon tubes are the flattened horseshoe-shaped evacuated tubes that were made by Eugene Bourdon (1808–1884). Perhaps these were available first and Vidi aneroids soon followed: Vidi aneroids used the corrugated disc type of evacuated capsule which became more popular with manufacturers. Either way, we can assume that aneroid-type barographs were available by 1868. I suspect that some may have appeared soon after Vidi's aneroid patent ran out around 1852 after the death of E. J. Dent, but they certainly would not have been common.

After mercury self-recording instruments, the aneroid type must have seemed a natural step forward. From the evidence so far, though, it appears not to have been developed until 1867 by Breguet (see chapter 2) and it is in 1868 that the first self-recording patent appears in Britain. Considering the price difference between the imported French barographs made by Richard Frères (see chapter 2), which were nearly a quarter of the price and were available from 1880, it is likely that far fewer were sold after 1885. This theory is supported by the fact that very few of the large self-recording

Fig. 1.15 Internal parts of a Negretti & Zambra self-recording barometer, c.1875.

aneroid barometers appear in auctions; although some undoubtedly were made later, the numbers would have been small. *Fig.* 1.15 shows the inside of a Negretti & Zambra oak self-recording aneroid barometer, c.1875. The large wheel fixed to the double capsule aneroid movement transmits the movement in pressure to the pencil holder on the right of the chart. The quality of the mechanism used is typical of these large self-recording barometers, but it is plain to see why Richard Frères' new instruments took over the British and probably world market for several years.

Fig. 1.16 shows another similar self-recording aneroid barometer from an 1880s' catalogue of James J. Hicks. The price for this one with the carved top is £17 17s, considerably cheaper than Negretti & Zambra's model, but for what reason we can only imagine. Negretti & Zambra's 1888 encyclopaedic catalogue illustrates the same engraving of a self-recording aneroid barometer still priced at £22, although the more decorative model in a carved case was £27 10s. Ruled charts were £1 1s per hundred. This catalogue also has Richard Frères' self-recording aneroid barometers, although advertised as theirs!

Fig. 1.17 shows a page of a Negretti & Zambra catalogue which was over-printed with 1920 prices. The late Victorian-style, probably ebonised, instrument shown was still being advertised alongside the now more traditional style. Interestingly, no new price was shown for the self-recording barometer illustrated – the old price of £27 was just ruled out and not

Fig. 1.16 A self-recording aneroid barometer illustrated in an 1880s' catalogue of James J. Hicks. © National Museums Scotland.

amended. Perhaps they no longer sold them and were using a pre-First World War catalogue. The other instruments advertised had doubled in price so we can assume that the price would have been £54 or more if available.

This catalogue mentions the word 'barograph' many times, which was by now a name in common use, but the catalogue still uses the term 'self-recording'. Soon after this date, only 'barographs' are mentioned. In L. Casella's catalogue, with an introduction by Casella dated 1871, the word 'barograph' is mentioned several times in relation to 'King's Barograph' and the 'Kew Barograph'. This is the earliest record I have found for the use of the word 'barograph'. Casella may have been the first to use the word, but it is hard to say how quickly or generally it was adopted. In the patent by George Meyer and Antoine Rédier of Paris, no. 7323 of 1891, no mention of 'barograph' is made. However, in the patent by Joseph Bartlett of Clerkenwell, London, no. 1900 of 1894, the complete specification left on 29 October 1894 states: 'This invention relates to instruments (of the character known as "barographs")'. In a catalogue by Elliott Bros dated 1902 a simple barograph is illustrated but, interestingly, it is entitled 'Recording Aneroid (Barographs)'. The obvious bracketing of the word 'barograph' indicates that it was still not the commonly known name for these instruments.

Fig. 1.18 shows a fine figured walnut self-recording barometer by Dol-

Fig. 1.17 Ebonised self-recording barometer as illustrated in Negretti & Zambra's catalogue in use in 1920.

Fig. 1.18 Walnut-cased self-recording barometer by Dollond of London, c.1880.

Fig. 1.19 Oak cased self-recording aneroid by Negretti & Zambra, 1876.

Fig. 1.20 Ebonised cased self-registering aneroid by R. & J. Beck of London, 1870.

lond of London, c.1880. It is complete with a minimum/maximum thermometer underneath the recording drum. When the front glazed door is opened downwards, there is often revealed a fine wire stapled into position. It is used while changing the chart to keep the pencil out of the way. The loose end of the wire, which is very springy, is lifted up and located onto the pencil carrier to keep some positive tension on it whilst the chart is changed.

Fig. 1.21 Original sales invoice for self-registering aneroid shown in *Fig.* 1.20.

Fig. 1.19 shows another example of a self-recording aneroid in a large oak case by Negretti & Zambra. This one measures 26 inches wide by 7⅛ inches deep by 16 inches high (65 x 18 x 40 cm); the dials are silvered with inner decorative gilded brass centres. This one appears to be dated 1876. It has a minimum and maximum thermometer mounted below the recording drum, and the front glazed panel is hinged down from the top to gain access.

Although most early barographs of this design are of the same basic shape, the one illustrated in *Fig.* 1.20 is another style occasionally found. It is probably less visually attractive, so no doubt went out of fashion. It is by R. & J. Beck, 68 Cornhill, London. The ebonised case, with hinged glazed door to the front, measures 17 inches wide by 8½ inches deep by 23 inches high (43 x 22 x 58 cm). Extremely rare is the original sales invoice (*Fig.* 1.21), dated 28 July 1870, which shows that it was sold for £25 and was called a self-registering aneroid barometer.

2 Aneroid Barographs pre-1902

The Richard family from France was responsible for the development of the traditional barograph, as it became known. The French inventor Eugene Bourdon (1808–1884) sold his part of the patent relating to barometers to Felix Richard around 1850. Richard was a successful instrument maker who later suffered financial ruin. His two sons, Jules Nicolas Richard and Max-Felix Richard, exhibited with their widowed mother in 1878 at the Paris Universal Exhibition, and in 1880 Jules patented a number of improvements for recording barometers. Drawn in this patent are details of fixing charts on drums with sprung clips and detailed drawings of the nib to hold ink.

It appears that it was the brothers who were the first to introduce these recording instruments. The number of early barographs still surviving that bear their design, if not their actual initials, is strong evidence. In 1882, the cooperation of the two brothers was formalised and they founded the Société Richard Frères (Richard Frères meaning Richard Brothers in French). After this date, the trademark RF appears on almost all their instruments. The year 1891 marked the end of the partnership, which then became Jules Richard, but the business kept the initials 'RF' so it is difficult to date actual barographs if only 'RF' is marked on them (see Brenni 1996b, p. 11).

In Negretti & Zambra's catalogue of 1885 a Richard Frères recording barometer can clearly be seen offered for sale at £7 10s (*Fig.* 2.1). Unmistakable proof that Negretti & Zambra were buying from France is on the page opposite in my copy, which has a cutting from a French catalogue (not proved to be Richard Frères but it is French and their design). This cutting is of a recording hygrometer wet and dry bulb type instrument; it has prices marked in pen of £1 1s and £12 12s which bear no relevance to the item but could be the purchase price and the selling price. Some explanation about my copy of the Negretti & Zambra catalogue would probably help the reader to understand some of my assumptions. It is the catalogue used by Paul Ernest Negretti, a grandson of the founder Henry Negretti, during his time at this famous firm. In it are numerous hand-written details about various articles, including details about items bought and for how much they were sold. It was obviously used regularly as reflected in its present condition. It came to me via one of their employees, Jack Noble, with whom I have had many conversations about Negretti & Zambra. He worked for them for most of his life and took an interest in their history.

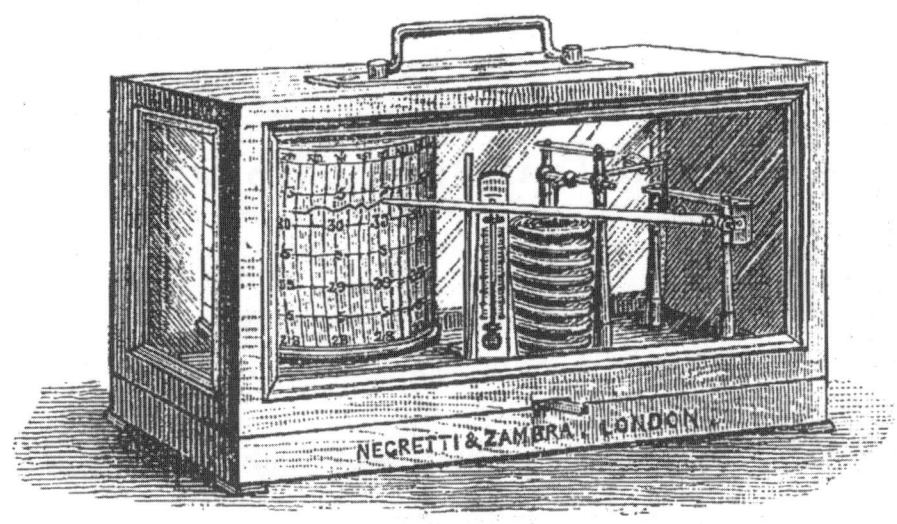

Fig. 2.1 Self-recording aneroid barometer by Richard Frères in Negretti & Zambra catalogue of 1885.

A catalogue by Louis P. Casella dated 1894 was brought to my attention by a collector. It is an illustrative and descriptive catalogue of recording instruments by Richard Frères of Paris by L. Casella of 147 Holborn Bars, London EC. In an address inside the cover, Louis Casella states that he has been appointed agent for instruments made by Richard Frères in the British Isles, India and the Colonies. More interestingly, he writes the following passage:

> Between 1840 and 1880 many methods for obtaining continuous records were devised; in most of them photography was employed, involving (1) the maintenance of a source of light, (2) manipulative skill in fixing the photograph, (3) the impossibility of examining the record while the phenomena were occurring, (4) bulky apparatus, and (5) considerable cost. All these difficulties have been swept away by the various patterns of apparatus described in the following pages, the appreciation of which is perhaps best illustrated by more than 15000 of these instruments having been sold since 1882, and by the fact that there is hardly an observatory in the world where some of them are not at work. They are in operation on the Eiffel Tower, and in coalmines, and for industrial

purposes too numerous to quote.

Success such as this has naturally led to the manufacture and sale of imitations. All genuine ones bear the Trade Mark R.F.

I can well believe that copies and similar styles were being made in Britain by the 1890s. As Casella does not say he is the sole agent, perhaps there were others selling these items, or firms such as Negretti & Zambra were perhaps buying from Casella. His assertion that 'many methods for obtaining continuous records were devised' is likely to have been correct as a number of patents have come to light, as well as other documents such as a paper on photographic self-registering instruments, together with drawings of a sizeable piece of equipment, read to the Royal Society by fellow Francis Ronalds on 21 January 1847 (see Ronalds 1847; National Meteorological Office Library and Archive, ref. 733-6-1985).

On page 7 of the Casella catalogue is a description and illustration of 'marine suspension' which was by then compulsory on French warships. This is not a style of suspension that I have come across on British ships despite looking for any tell-tale signs of wear under the handle from movement on a ship which might be expected if used for some time aboard. *Fig. 2.2* shows Short & Mason's barograph leaflet, regrettably undated, as most are, but it looks to me as if it could be from the 1950s. In the middle is exactly the same suspension hook for barographs at sea as Louis Casella was advertising in 1894, so they were possibly used in this country as well as in France. Casella's price for a full-size barograph was £5 10s, the smaller one was £4 10s and the same with glass on three sides was £5 15s – somewhat cheaper than Negretti & Zambra's catalogue prices of 1885.

Further evidence that Richard Frères were leading suppliers to Britain is, I believe, the lack of British patents during this time of development. There are a number of patents, as can be seen in the list of patents in the Appendix, but none really has any serious impact or importance until 1902 when Short & Mason, who were working from 1873 onwards, patented their design for mounting the clock and a different chart fixing. Short & Mason and Wilson Warden & Co. appear to have been major producers of barographs during the first half of the nineteenth century. It is obvious from handling many barographs over the years that they did not always mark their product with their own name but made them for others to sell, a very common practice among many instrument makers. Both these firms were of long standing and had a reputation for good quality, yet it is not until the 1920s that evidence for their manufacture of barographs becomes apparent from existing examples being signed by either of them. No doubt there were other British makers, but I believe that British-made barographs did not appear generally until the 1890s.

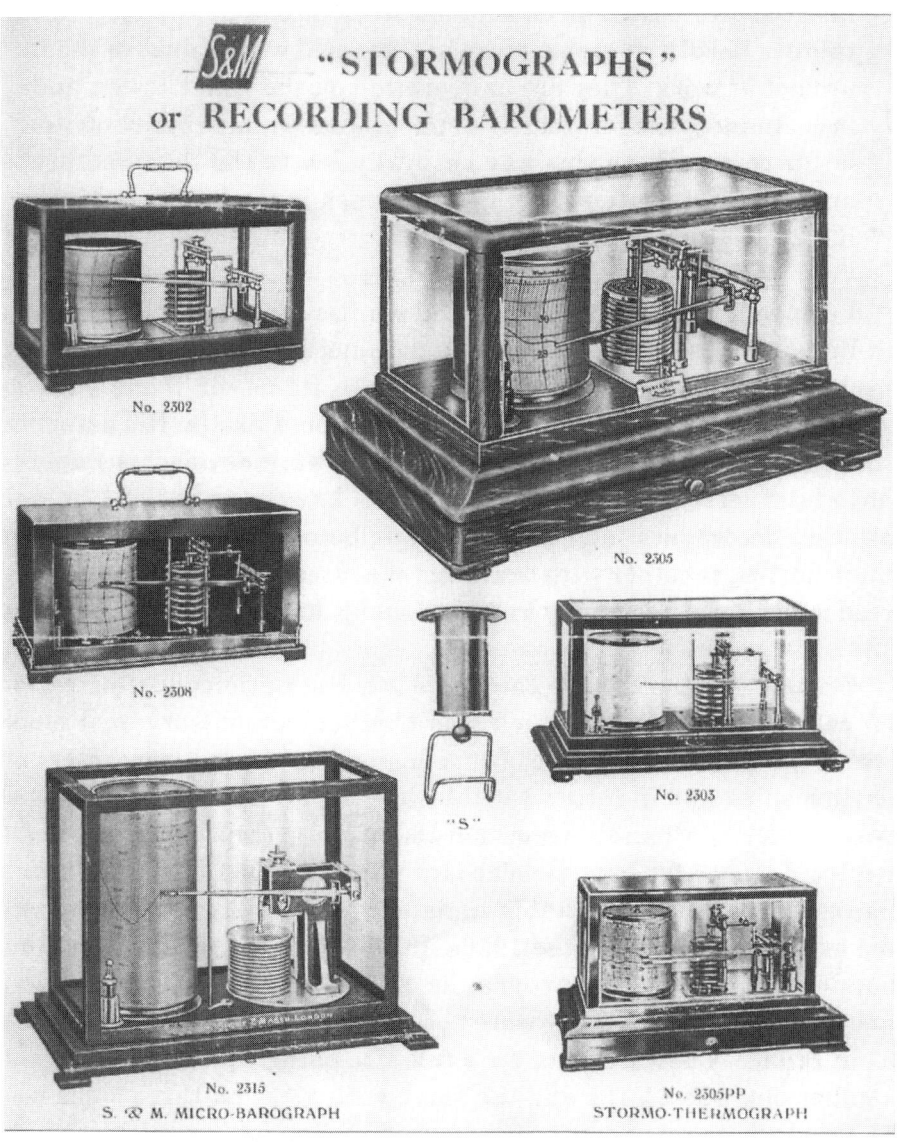

Fig. 2.2 Barograph sales leaflet by Short & Mason dating probably from the 1950s.

Fig. 2.3 shows a typical French barograph by Richard Frères, numbered 5669, and likely to be reasonably early, perhaps c.1885. The machine-cut joints on the case, visible in the illustration, once led me to think that these were considerably later, but I am now of the opinion that these date back to 1880 when the patents were registered. The handle appears to be of

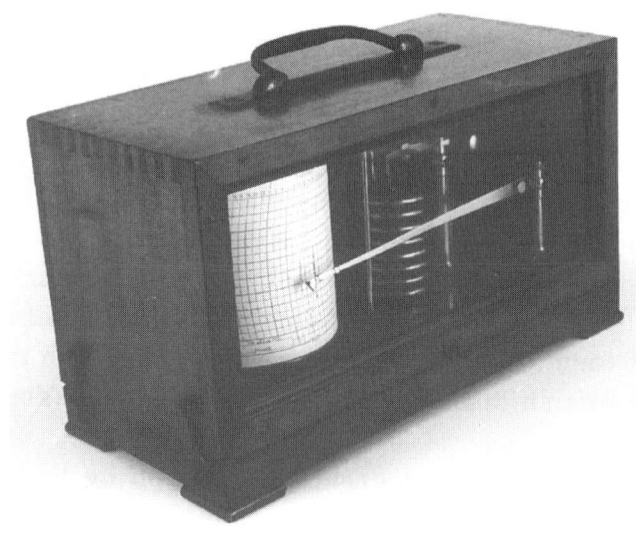

Fig. 2.3 Typical early barograph by Richard Frères, number 5669, c.1885.

Fig. 2.4 Richard Frères movement (*Fig.* 2.3), showing throw-off arm and protective bracket behind the arm.

Fig. 2.5 Double-ended key for Richard Frères barographs.

steel; the hooks, which are usually two at one end and one or two at the other end, are normally of steel, although sometimes they are plated. The steel ones are probably earlier. The characteristic throw-off arm, with the brass coming through the front of the case, and the mechanism, which can be seen in *Fig.* 2.4, are of a fairly standard design. Richard Frères' models have a polished steel bracket protecting the arm from the case when it is removed or replaced. This is a very sensible precaution as it is so easy to knock the arm and splash ink about the case or mechanism when removing or replacing the lid. I often wonder why later manufacturers did not do the same.

The adjustment for altitude is almost hidden beneath the barograph. There is a small hole in the wooden case in which to insert the smaller square of the key (see *Fig.* 2.5) to locate on the square end of the adjustment screw. Turning it left or right adjusts the tension on a heavy brass plate, which has the bellows stack attached, correspondingly to alter the reading for altitude adjustment. These barographs nearly always have a double-ended key supplied with them, one end for winding the clock and a much smaller square end for inserting beneath the barograph to adjust the tension on the bellows stack. The arm is of the usual tension type. The brass-work is generally very finely turned.

It is interesting to compare British-made barographs to these early French ones: even some of the more modern ones made into the 1960s have the same design of pillars in many cases. This is almost certainly due to the fact that Richard Frères cornered the market in the early days and most people have copied this design through the decades. Of course, there are exceptions to this, such as the Casella type with simple plain pillars and the barographs made by Gluck Engineering (see *Fig.* 2.6). Gluck used automatic type lathes to mass produce parts, and the design of the pillars (as well as other parts) was altered to make turning more suited to such manufacturing. My own design of barograph (*Fig.* 2.7) incorporates a number of features from old barographs and especially the traditional shape of the pillars.

The mahogany case of the Richard Frères barograph (see *Fig.* 2.3) has one glass panel to the front, which is usual; the mitred moulding to the front is screwed into the case with brass screws visible. Four small flat pad feet are also the norm for this style of barograph. It is very similar in design to the later Meteorological Office fisheries barographs (see *Figs* 3.31 and 3.32 in chapter 3). Probably for this reason they have often been passed over as not being as interesting as the full display models that the British makers so often produced.

Another French barograph is illustrated in *Fig.* 2.8. It is numbered 73013 and inscribed 'Richard Frères, Brevetés, SGDG, Paris': the trademark can be seen in *Fig.* 2.9. The dimensions of the barograph are 12 inches by 6½

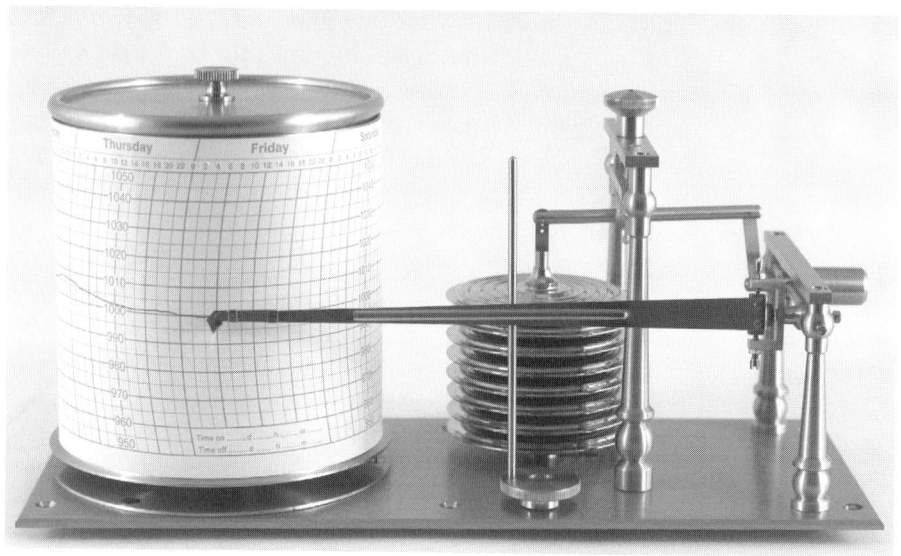

Fig. 2.6 Barograph movement by Gluck Engineering.

Fig. 2.7 Modern barograph by Collins of Merton.

Aneroid Barographs pre-1902 33

Fig. 2.8 Brass-cased barograph by Richard Frères, c.1900.

Fig. 2.9 Trademark of Richard Frères barograph (*Fig.* 2.8).

inches by 6 inches high (30 x 16.5 x 15 cm); it is completely brass cased with five bevel-edged glasses and a traditional design movement, dating to c.1900. The clock has a paper-retaining clip and is secured in position by a knurled knob. There is the usual right-angled steel protection behind the arm to protect against knocks when moving the lid. The thermometer (in centigrade) is mounted between the charts and the bellows stack. Originally ormolu gilded, now in gold lacquer finish, the four screws retaining the two main pivots are not slotted but have two flat sides for a spanner or a pair of pliers to adjust. This is a good idea as more leverage can be had than with a slotted screw, which in time becomes stuck or corroded. The throw-off arm has a small knob to help to move it – a slight modification, I suggest, from the style that extends beyond the case (as can be seen in *Fig.* 2.12, for example). The base comprises a flat brass bed screwed to the cast moulded frame, which has four feet, one screwed to each corner. 'SGDG' probably stands for the society that Richard was a member of, although I have not yet been able to confirm this. According to Maxant (2000, p. 176), Felix Richard founded a society for the construction of barometers in 1845.

Fig. 2.10 is a mahogany barograph by Richard Frères, number 15257, c.1890, measuring 11½ inches by 5⅜ inches by 6½ inches high (29 x 13.5 x 16.5 cm). It is a typical design from this maker, with three glazed sides. This poor example has seen some considerable use or misuse. It has had brass-covered straps added to strengthen the joints and a single small hinge and clasp at one end to make lifting the lid easier. The three hooks and eyes have been dispensed with. No doubt some well-meaning restorer has given the whole woodwork a liberal coat of varnish, but fortunately the colour has survived well. A much more modern plate has been nailed in front of the base plate, which states that it is 'MADE IN FRANCE'. The carrying handle shows much pitting but no signs of the chrome or nickel-plating that is normally expected. There is some type of black paint remaining in the nooks and crannies, and the screws that hold the lid together are original and quite typical of this style.

Fig. 2.11 shows a mahogany barograph of around 1890 with four glazed sides and a label inscribed 'James Lucking, Optician, 5 Corporation Street, Birmingham'. It has a hinged lid and heavy brass late-Victorian carrying handle. There is an unusually narrow drawer in the base for spare charts. As with so many of these styles, the lid has suffered over the years, the glass has been replaced and pinned in with tacks and one end has been puttied in. There are only small remains of the original fine mahogany beading that usually holds the glass in from the inside. The hinged end has been strengthened later, probably due to breakage. Always be careful when carrying these barographs. Although designed for carriage, I never like to use the handle alone, but carry them with both hands underneath – just in case!

This example is a good barograph in need of some cleaning and res-

Aneroid Barographs pre-1902

Fig. 2.10 Richard Frères barograph number 15257, c.1890.

Fig. 2.11 Mahogany barograph inscribed James Lucking, c.1890, with thermometer.

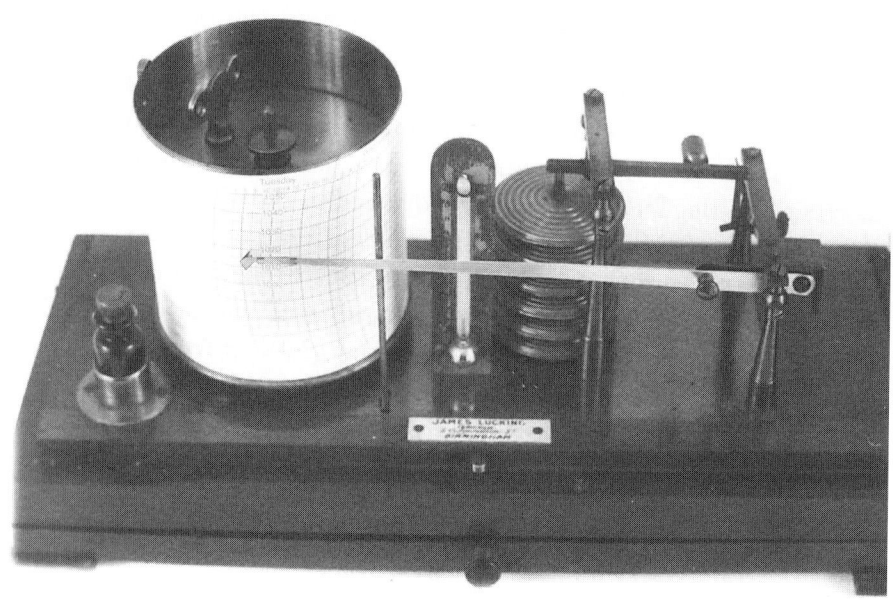

Fig. 2.12 Mechanism of barograph in *Fig.* 2.11, showing usual throw-off arm.

toration. The thermometer scale is badly corroded and would benefit from re-silvering and lacquering to preserve. However, the owner of this item only wanted it to work, so it left the workshop looking no better but operationally fine. Interestingly, the stack of capsules is side flanged with soldered nipples for evacuating them. This is an obvious replacement as these only came in from 1922 onwards. When received, it was not moving enough so we dismantled it and adjusted the lever positions whilst checking in a pressure chamber. The knurled knob on the right near the hinge end, which can just be seen in *Fig.* 2.12, is the fine tuner for making the barograph more or less sensitive. It moves the lever connecting to the pen spindle in or out, similar to some earlier aneroid movements. The throw-off arm is moved by a short lever sticking out under the main plate but not protruding beyond the lid. One therefore needs to open the lid first to throw the pen away from the chart. In some models, like the Meteorological Office fisheries barograph, this lever is accessible outside the closed case, a minor difference. The polished steel bracket protecting the end of the tensioned arm is a clue to the fact that it was actually made by Richard Frères; the adjustment for altitude is again underneath, as in Richard Frères examples. This was certainly made by them and sold by James Lucking.

Another example of this type of 'badging' is shown in *Fig.* 2.13, clearly another Richard Frères barograph but sold by Elliott Brothers. It is in a rather sad state: being uneconomical to restore, it was left with us by its owner. The case is very basic: it has two brass brackets on the left and right with holes in the horizontal part which were designed to screw the item down securely perhaps to a desk on a ship. It certainly is an early barograph and has five glass panels. The clock is also of French manufacture as shown underneath (see *Fig.* 2.14). A close examination of the underside of the base plate (*Fig.* 2.15) shows a number of long thin dents: this is where Elliott Brothers' engraver hammered out the engraved details of Richard Frères to enable them to engrave their own name. The brass plate was thick enough and soft enough to take this amount of hammering, after which the plate would have needed considerable smoothing to finish up with a flat, new-looking plate ready to engrave.

Fig. 2.16 is a small, mahogany-cased barograph with a half-sized clock drum, chart drawer and an overall size of 11 inches by 7¾ inches by 6½ inches high (28 x 20 x 16.5 cm), sold by Negretti & Zambra of London. The bevelled-edge mahogany case lid is held in position by two brass screws. The fine adjustment for calibration can be seen between the short pillars and is by means of a knurled nut lengthening or shortening the lever connecting it to the rod link. The main adjustment is underneath by using a small key to change the height of the bellows stack, the same design as many barographs by Richard Frères. This is a very attractive barograph, probably dating from 1890. The throw-off lever for the pen arm is another feature by Richard Frères but cannot be used without taking off the glazed lid. It is so similar to the barograph shown in *Fig*s 3.42 and 3.43 (see chapter 3) that it can be assumed to be by Richard Frères.

A notable British firm in instrument making, particularly relating to mining, is Davis of Derby. It was started by Gabriel Davis in 1779, and although the Davis family sold the business in 1963, the company continues to flourish, nowadays predominantly in electrical and electronic equipment. The company was for a long time known as John Davis & Son, Derby Limited. *Fig.* 2.17 is from their catalogue number 343A. The catalogue probably dates from the early twentieth century, featuring, as it does, a three-figure telephone number. The illustration is quite likely, as with many catalogues, to be from an earlier period. The arm of the recording instrument appears to be narrower and possibly stiffer than the aluminium-type arm, but this could be the illustrator's interpretation. The bellows stack is of the soldered sides variety and there is an interesting cut-glass ink bottle. You will notice the metal-tap affair on the right-hand side. This is not, in fact, a barograph but, as their catalogue states, a 'barograph type'. It is a self-registering water gauge and the capsules are not evacuated but operate recording water level. On a number of occasions, I have had these brought in to the workshop as barographs and have had to disappoint the usually recent purchaser with

Fig. 2.13 An early barograph made by Richard Frères and sold by Elliott Brothers.

Fig. 2.14 Detail under clock of *Fig.* 2.13.

Fig. 2.15 Hammer marks to remove the original maker's name on the other side of the plate.

Fig. 2.16 Small barograph inscribed Negretti & Zambra, c.1890.

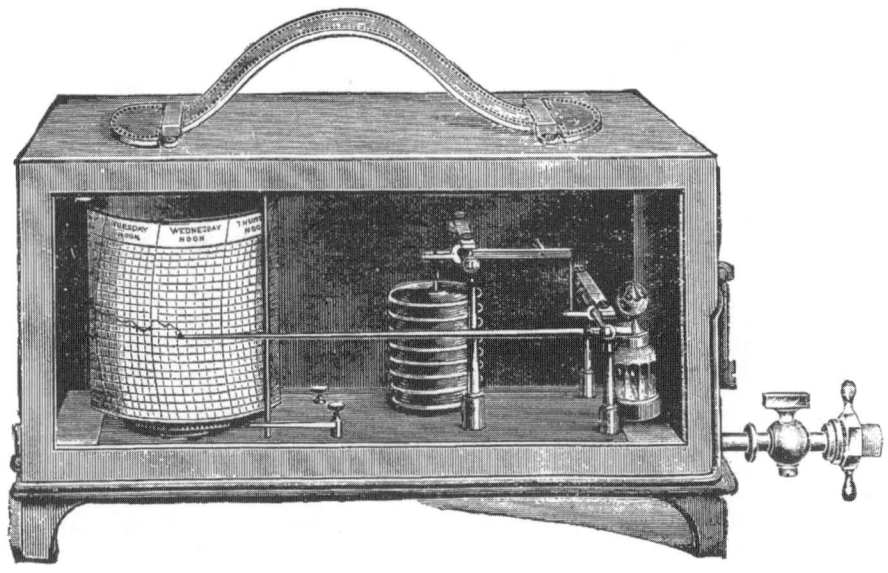

Fig. 2.17 Early twentieth-century self-registering water gauge by Davis of Derby.

the news that their instrument is not, in fact, a barograph. We have successfully converted one to a barograph using modern barograph bellows and altering the linkages and levers. They usually have a cast-iron base with a copper cover and were made for a number of years as an industrial instrument.

Fig. 2.18 shows two barographs offered for sale by Davis. Both have side-soldered edges to the capsules, and were available in different woods or ebonised. These two particular barographs probably date from around 1907. The pen arm still retains the right-angled bracket to protect it from being knocked by the case when opening and closing. The lower of the two is a barograph with a barometer dial, which was offered for sale at £6 10s. A close inspection of the dial reveals the logo of Short & Mason (S and M pierced by an arrow) and the registered number 628606, proving again that, although Davis were fine instrument makers in their own right, in this instance the barographs were made by another company. I believe the top illustration is of an earlier drawing.

In *Fig.* 2.19 we see a heavily carved and well-constructed oak barograph with subsidiary dial by J. Casartelli & Son of Manchester. This firm is particularly known for making siphon tube barographs. The one illustrated here probably dates from 1890. It is a fine example. Later dial barographs generally have open dials, so this closed dial is quite unusual.

Self-Registering Barometers.

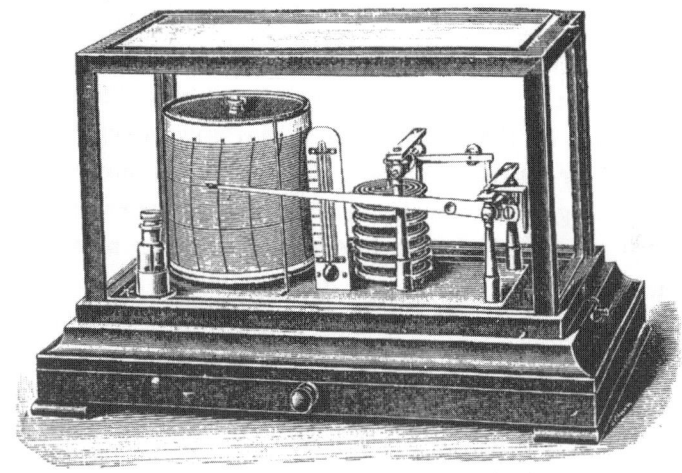

This instrument shows *at a glance* the fluctuations which have taken place in the atmospheric pressure. The Paper Chart attached to the drum revolves by clockwork, automatically records the variations which have transpired during the week, indicating the precise times of occurrence and periods of duration of each change.

The paper upon which the diagram is recorded should be removed every week.

No. 0123.—With Glass Shade Cover, ebonized base £5 8 0
 ,, 0323.—With wood frame in mahogany, oak, walnut, or ebonized, and glass all round and top, with thermometer, as illustrated £6 0 0

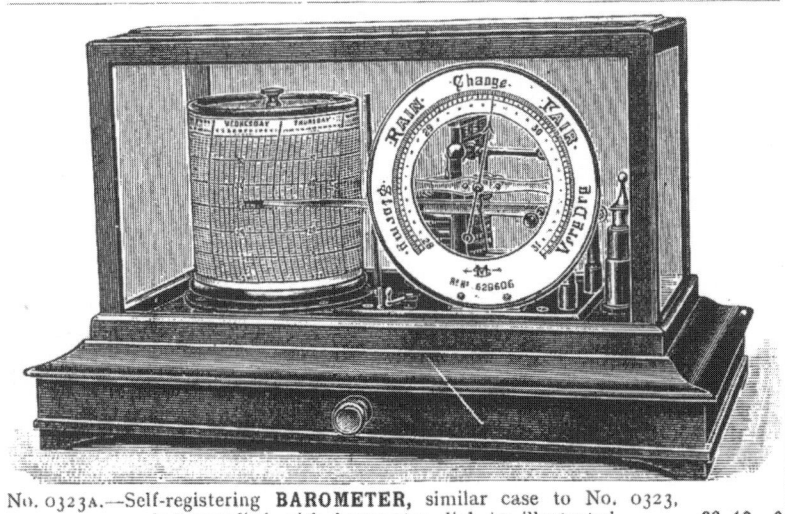

No. 0323A.—Self-registering **BAROMETER**, similar case to No. 0323, but supplied with barometer dial, as illustrated .. £6 10 0

No. 3423.—Self-registering **THERMOMETER**, 20° below zero to 115° above, Fahr., with metal case in copper, with handle .. £7 0 0

Fig. 2.18 Barographs offered for sale by Davis of Derby, c.1907.

Fig. 2.19 Carved oak barograph by J. Casartelli & Son, Manchester, c.1890.

Amongst the finest examples of barographs handled in our workshop is the one illustrated in *Fig.* 2.20, an inlaid mahogany barograph by Pillischer of 88 New Bond Street, London. The case is well inlaid with coloured wood and stringing. It measures 16½ inches by 11 inches by 10 inches high (42 x 28 x 25 cm). According to Edwin Banfield in *Barometer Makers and Retailers* (1991, p.170), Moritz Pillischer was working from 1850 to1888, the latter part of this time at New Bond Street, where he was followed by his son, Jacob Pillischer. This barograph is almost certainly by the son, Jacob Pillischer, and can reasonably be dated to around 1900. The arm is of the sprung type and has its adjustment by a screw rather than a knurled knob, which indicates that the adjustment was not frequently considered necessary or perhaps it is a later replacement.

Fig. 2.21 shows the instrument with the top removed and the clock drum displaying a very unusual clock arrangement. The clock is wound centrally as opposed to the more usual off-centred winding, the key normally staying in position on a left-handed thread under the drum lid. The clock movement is lifted up on a platform, and although in principle similar to patent 3715 of 1902, this could be perhaps a competitor's variation or just a coincidental change. However, it certainly means that the drive would be towards the middle of the drum as opposed to the earlier models, which were usually driven at the base of the drum. A thermometer is mounted centrally to the front of the plate and an engraved name can be seen just to

Aneroid Barographs pre-1902

Fig. 2.20 Inlaid mahogany barograph by Pillischer, c.1900.

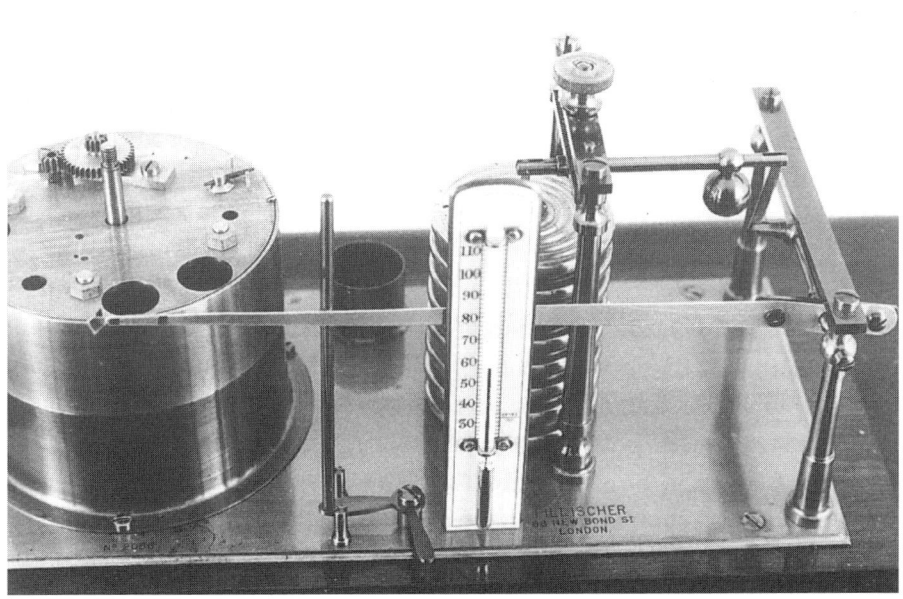

Fig. 2.21 Unusual clock movement of the Pillischer barograph (*Fig.* 2.20).

the right of this. The pillars are of a slightly different design from usual, the adjustment for altitude being by the knurled screw above the capsules. A close-up view of this type of altitude adjustment is given in *Fig.* 2.22. It became almost standard on British barographs and always on Short & Mason instruments, although the Richard Frères design was much simpler to make.

At first glance, *Fig.* 2.23 looks like an ordinary oak-cased barograph and probably dates from around 1890. It has bevel-edged glass and measures 14¾ inches by 8¾ inches by 8½ inches high (37 x 22 x 21 cm) on four moulded pad feet. The altitude setting adjustment is in the base plate between the two pairs of pillars. The retaining nut on the clock is missing, although these retaining nuts are really only of use for transit, as in general use the weight of the clock sits satisfactorily on the gears. It is an un-named barograph with no clue or indication as to manufacture. The clock is of pre-1902 design and the arm is of the sprung type. However, as can be seen in *Fig.* 2.23, the arm is almost fully visible and, if comparing with the top-down view shown in *Fig.* 2.24, it can be seen that the arm is mounted on a bracket and comes forward of the pillar rather than being tucked behind it.

This is an unusual variation that I have not come across before and makes some operations far easier as the spindles do not have to be removed, as in the usual type, to replace the arm if needed or to detach it to clean the nib. The balance weight, fixed to the connecting rod from the bellows, which reduces the slop in the pivots and thus maintains slight positive pressure on all the linkages, whether the pressure is rising or falling, is of much larger design and is also adjustable. There is a screw in the top of it to locate it onto the arm in the best position to be most effective. A third notable variation is the throw-off arm, which is of slightly different design, being more of a 'V' shape, as can be seen in *Fig.* 2.24.

Fig. 2.25 shows a golden oak-cased barograph on four moulded pad feet, with flat glass, attributed to Yates & Son of Dublin. The barograph and clock are held on one long, blued brass base plate; the pillars and remainder are of brass in gold-plated finish and the steel screws are all blued. The combination of these complementing colours on the metal work makes for a very pleasant appearance. The arm is of the tension type; the altitude adjustment is by means of a knurled nut above the bellows stack. The ink bottle, 1¼ inches (3 cm) in diameter, is original; although not cut-glass, it has a ground stopper. The clock is of the pre-1902 design and probably dates from around 1895. The most saleable and sought-after barographs are usually those with a drawer and bevel-edged glass. Many collectors and dealers tend not to value barographs so much without a drawer, though many simpler and earlier good-quality barographs, like this one, can often be very accurate instruments and may well represent a good opportunity to acquire what is an older and sometimes more interesting barograph.

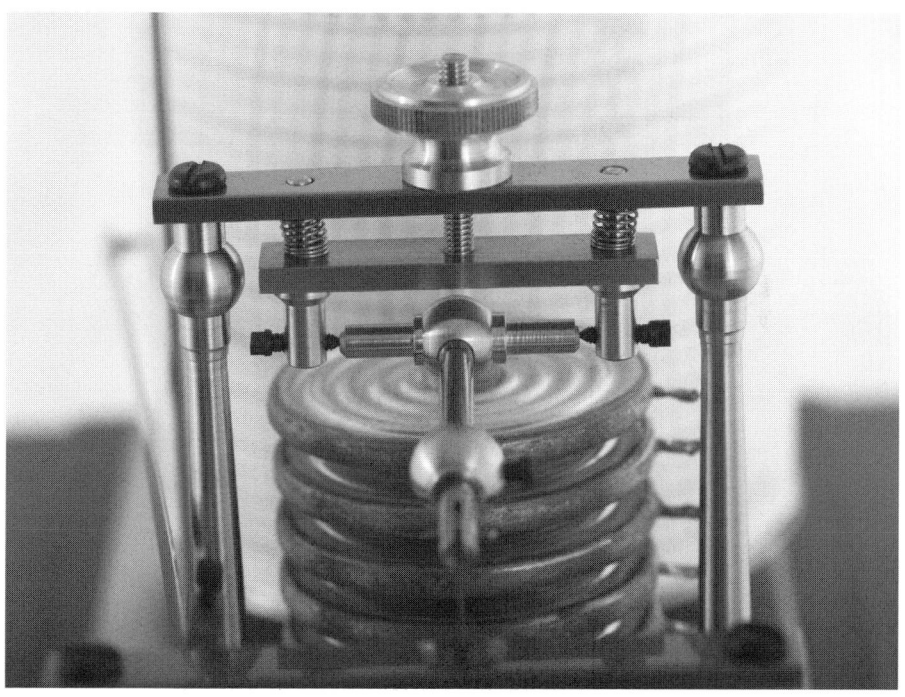

Fig. 2.22 Close-up view of the type of altitude adjustment used in the Pillischer barograph (*Fig.* 2.20).

Fig. 2.23 Oak-cased barograph, c.1890.

Fig. 2.24 Top-down view of *Fig.* 2.23 without glass, showing unusual arm arrangement.

Fig. 2.26 illustrates another simple, golden-coloured oak-cased barograph with flat glass, this time by T. B. Winter & Son, 21 Grey Street, Newcastle upon Tyne. The case has moulded pad feet; the altitude adjustment is on top of the four pillars, the arm of sprung design and the clock of the pre-1902 type, the mounting of which can be seen in *Fig.* 2.27 with the clock removed. *Fig.* 2.28 shows the top of the drum with the lid removed, with the key in position and a dust cover to the fast/slow adjustment. This barograph probably dates from around 1890. T. B. Winter & Son sold numerous models, which always appear of good quality.

Fig. 2.29 is a mahogany barograph inlaid with boxwood stringing and lines to the edge of the removable case, drawer front and around the base, also by T. B. Winter & Son, 21 Grey Street, Newcastle upon Tyne. *Fig.* 2.30 shows the traditional design mechanism with older-style clock mounting and tension arm, suggesting a date before 1902. This case originally had a lift-off lid, but at some later stage it has been fitted with a hinged lid. The drawer is mahogany framed, the base sitting on small bracket feet. It is a desirable barograph: it has a nice colour polish and the charts are held on with a hinged clip.

Fig. 2.31 is an oak-cased flat-glazed barograph by John Barker & Co. Ltd, Kensington, with oak drawer and chunky-looking case, probably c.1895, with sprung arm and twin-capsule mechanism of traditional aneroid style. The close-up view in *Fig.* 2.32 shows more clearly how the set hand pen

Aneroid Barographs pre-1902

Fig. 2.25 Golden oak-cased barograph by Yates & Son of Dublin, c.1895.

Fig. 2.26 Oak-cased barograph by T. B. Winter & Son, c.1890.

Fig. 2.27 The movement of barograph by T. B. Winter & Son (*Fig.* 2.26) with clock removed.

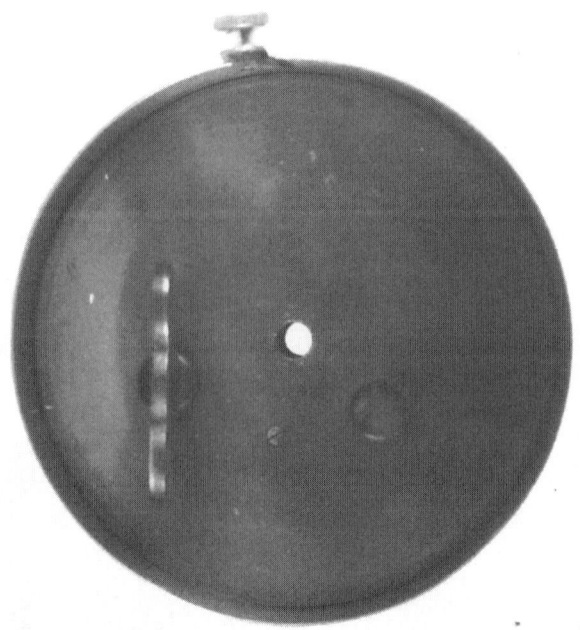

Fig. 2.28 Clock drum of barograph by T. B. Winter & Son (*Fig.* 2.26) with lid removed.

Aneroid Barographs pre-1902

Fig. 2.29 Inlaid mahogany barograph by T. B. Winter & Son, c.1900.

Fig. 2.30 Mechanism of barograph by T. B. Winter & Son (*Fig.* 2.29).

Fig. 2.31 Oak-cased barograph by John Barker & Co. Ltd, Kensington, c.1895.

Fig. 2.32 Mechanism of barograph by John Barker & Co. Ltd (*Fig.* 2.31).

is screwed through the L-shaped arm of the mechanism and adjusts the tension on the C-spring, the C-spring being nickel-plated, the fulcrum arm fixed low to the base plate as would be found on aneroids. Although some barographs of this design can work well, I always consider it a cheaper version than the more usual stack of capsules. Over many years of restoring barographs, I have found that it is often this design that has problems and complications, and would avoid purchasing a barograph of this type unless the seller can prove it is working well. Many times, customers have brought into our workshops a barograph recently purchased from some dealer only to find that it does not work properly. Unless the price is good and you have confidence in buying, I would advise the novice to stay clear of this type of barograph.

Fig. 2.33 illustrates a large oak-cased barograph of unknown make with brass drop handle, which suggests a style of very late Victorian or Edwardian manufacture, with large flat bracket feet and double-stepped base above the drawer. This model, having had a replacement bellows of a modern movement, is over-sensitive and over-reacts to air pressure compared to other barometers due to the fact that the leverages on the old barograph were made for a less-sensitive bellows. The only remedy is to re-drill the base plate – thus spoiling the appearance – so as to re-position the linkages and leverages of the original mechanism. This would not normally be acceptable as it would tend to spoil the original nature of the brass work. Old-style replacement bellows, however, are now available which means that there is a better chance of replacing leaking bellows with ones that may not only work well but also look almost original (see chapter 6 for bellows).

Fig. 2.34 is a Victorian barograph of about 1890, with glass front and side-hinged lid, by J. H. Steward, 406 and 457 The Strand, London, which is nicely hand-engraved into the base plate. The decoratively turned brass setting knob is between the eight-capsule stack and the clock. The ink bottle is located into a brass loop, which is screwed to a fixed piece of wood below the hinge. Over the end of the arm near the hinge can be seen the polished steel bracket to protect the arm from impact when removing the lid. As this barograph is hinged, it is not necessary. This is a typical design used by Richard Frères, the French makers, and I believe in this instance can be attributed to copying their style. If it was their movement, I suspect that they would have used the underneath adjusting method which is so typical of their design. The decorative copper-plated handle and the hooks on the fixing side indicate that this may have been silver-plated but I cannot be sure.

Fig. 2.35 shows a late-Victorian ebonised cased barograph with mirror to the back of the case and bevel-edged glass to three sides and the top. The mechanism, unusually, is nearly all in polished aluminium, making a stark contrast to the black case. This was sold by the Army & Navy Co., West-

Fig. 2.33 Oak-cased barograph, un-named, c.1900.

Fig. 2.34 Hinged mahogany barograph by J. H. Steward, c.1890.

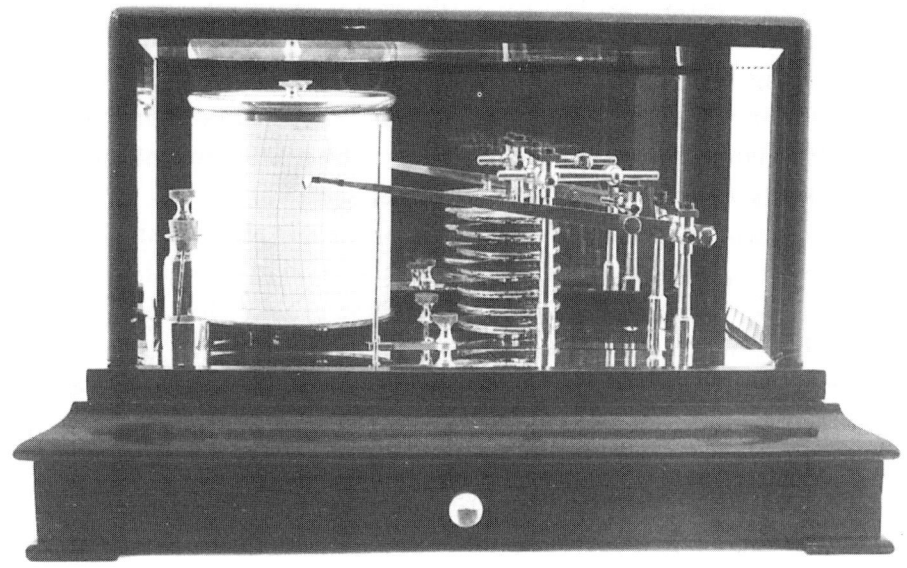

Fig. 2.35 Ebonised cased barograph with mirror back, c.1890.

Fig. 2.36 Walnut-cased barograph, c.1895.

minster, London, probably around 1890–1900, and is a very nice example of a slightly unusual barograph. *Fig.* 2.36 shows a walnut-cased barograph, c.1895, without a drawer, with eight capsules to the stack, and adjusting screw for the pen above the fulcrum arm; the glass is flat not bevelled. It is a slightly more economy model in terms of the case, but a very good mechanism with the sprung arm. The brass cup seen on the right of the mechanism would have held a large, possibly cut-glass, ink bottle and stopper; sadly, many of these have been lost as has happened with this one.

3 Aneroid Barographs post-1902

In chapters 1 and 2, we have seen a variety of efforts to make recording barographs from mercury barometers and the self-registering aneroid barometer. These developments led to what became established as the traditional barograph we recognise today. This chapter on post-1902 instruments will consider these traditional models, how similar they are and how it is probable that a few successful manufacturers tended to dominate the market – manufacturers such as Wilson Warden & Co. and Short & Mason.

The year 1902 serves as a useful dividing point to differentiate between styles of instrument because in 1902 Short & Mason filed a patent for a raised clock and glued-on charts. This was not a major breakthrough in itself, but these clocks are invariably numbered 3715.02 (being the number and year of the patent), and this serves as a clear demarcation between the earlier models and later ones. Naturally, not all manufacturers could use this patent, but it will become apparent from the large numbers that survive that this patent became the norm and strengthens the argument that Short & Mason were the major manufacturers of barographs during the early twentieth century. It is a design still in use today, although makers have generally reverted to the practice of securing the charts by a brass clip, similar to the Richard Frères original patent.

Short & Mason's patent was last renewed during the week 11–15 January 1915 and therefore expired on its fourteenth anniversary in 1916. During this time patents ran for 14 years only, provided that fees were paid. In brief, therefore, if you are considering buying a barograph and you see the numbers 3715.02 on the clock, you know that the clock at least is dated 1902 or later. If any dealer or auction catalogue describes it to you as a Victorian 1870 barograph – this is highly improbable.

It therefore seems likely that Short & Mason were major manufacturers of barographs, as indeed they were of aneroid barometers and no doubt other instruments, at the beginning of the twentieth century. The year 1902, at the start of the twentieth century and coming so soon after the death of Queen Victoria in 1901, seems an apt demarcation date for dating barographs. Dating is, of course, often difficult, but I have tried to date the barographs discussed here, according to the best available evidence, because I realise how important dates are to collectors and dealers alike. Some dates of course may need to be re-evaluated in the light of future research.

An American walnut-cased barograph is illustrated in *Fig*. 3.1, c.1905,

Fig. 3.1 American walnut-cased barograph, c.1905.

with Corinthian-type column corners, bearing simple incised carving, and with a Victorian-style handle to the drawer. It bears an ivory plaque inscribed Short & Mason of London, has a tension-type arm, an eight-capsule stack, 1902 raised patent clock and altitude adjustment from the knurled nut above the pillars over the bellows. The barograph measures 15 inches by 9 inches by 9¼ inches high (38 x 23 x 24 cm) on a moulded base mounted on bracket feet.

Fig. 3.2 shows an oak-cased barograph inscribed by Kelvin Bottomley & Baird Ltd, numbered D22130. The case measures 12¼ inches by 6¼ inches by 7½ inches high (31 x 16 x 19 cm). It has a gold-plated movement with an eight-flange-soldered bellows with evacuation spigots and tension-type arm, which is late for this style. The case has a nickel-plated carrying handle, small turned feet, a hinged lid and patented clock, number 3715.02. Overall, it is in excellent condition. A close-up of the mechanism can be seen in *Fig.* 3.3, almost certainly made by Wilson Warden & Co. It has a plaque on it recording that it was presented in May 1935, which goes to prove that not all barographs had the hinged gate-type arm after the 1920s.

Fig. 3.4 is an oak-cased dial barograph, c.1910, by Short & Mason, registration number 428606 (see patent list in the Appendix), measuring 14½ inches by 8½ inches by 8¼ inches high (37 x 22 x 21 cm) with five thick bevelled glass panels; the clock is of the 3715.02 patent but the drum uses

Aneroid Barographs post-1902

Fig. 3.2 Oak-cased barograph dated 1935.

Fig. 3.3 Close-up view of the movement of barograph in *Fig.* 3.2.

Fig. 3.4 Oak-cased dial barograph by Short & Mason, c.1910.

a retaining clip for the charts. It has a gold-plated movement with blued brass base plate and clock drum assembly. The tension arm is adjustable by the long knurled screw, visible at approximately the 30.8 division inside the open dial, as can be seen more clearly in *Fig.* 3.5, which also shows the hair spring arrangement to keep the hand under a little tension. *Fig* 3.6 shows the additional rod connected to the pen arm bar which connects to the arbour to which the barometer hand is fitted. Altitude adjustment is by the top knurled tension screw above the bellows stack. The case has a mahogany-lined drawer and simple bracket feet.

Fig. 3.7 is a simple oak-cased barograph, measuring 12½ inches by 7 inches by 8 inches high (32 x 18 x 20 cm). The cross bar is engraved 'patent 22556', which relates exactly to the patent of 1904 by Short & Mason (see patent list in the Appendix). At first glance, one could think that there may be a drawer beneath the front glass to the case but this extra spacing is required for the mechanism that is underneath. It comprises a standard aneroid movement with one of the links coming up through to the linkage arrangement on top. The top-down view (*Fig.* 3.8) shows this coming through a brass collet in the case. Above the ink-bottle holder is a similar hole with a brass collet and a knurled nut, which is the adjusting knob for altitude. The clock is of the 3715.02 design, so it is clearly after 1902 and

Fig. 3.5 Close-up view of the dial of barograph in *Fig.* 3.4.

Fig. 3.6 Detail of the linkage to the barometer hand of barograph in *Fig.* 3.4.

Fig. 3.7 Oak-cased barograph, c.1910, with Short & Mason patent.

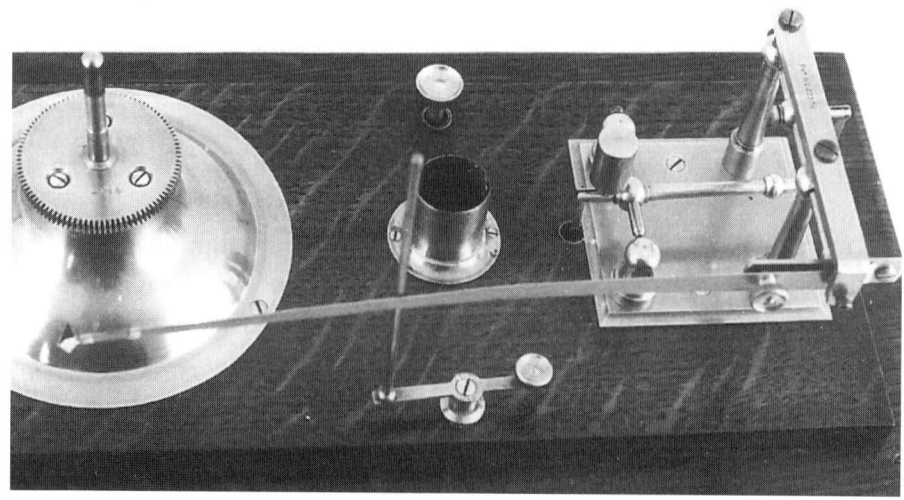

Fig. 3.8 Top-down view of barograph in *Fig.* 3.7 showing the hole where the rod connects to the movement under the slab.

still retains the tension arm type of pen, which is evidence that the hinged type was not yet in use. This instrument probably dates from around 1910 before the more standard Short & Mason styles evolved in the 1920s.

A sales leaflet on stormographs by Short & Mason with instructions for barographs contains the following: 'The S & M "Gate" pen arm in the form of a self-adjusting hinge eliminates friction between pen and chart. This feature is included in all S & M recording instruments.' This is a small pointer, I believe, to the possibility that Short & Mason patented or designed this hinge type of arm, but frustratingly the item is not dated so one cannot glean much further information from it. *Fig.* 3.9 shows a close-up view of this type of 'gate' arm which we use on our current barograph at Merton.

Fig. 3.10 illustrates an oak-cased barograph, measuring 14½ inches by 9¼ inches by 9 inches high (37 x 23.5 x 23 cm) with bevel-edged glass. The clock and drum are of the style often found on Meteorological Office instruments, the charts being held on by two sprung brass clips, which can be seen in *Fig.* 3.11. The capsules are of the flange-soldered type but still retaining the side nibs. The movement is also very similar to the Met. Office barograph, although this one has the adjustment for altitude knob between

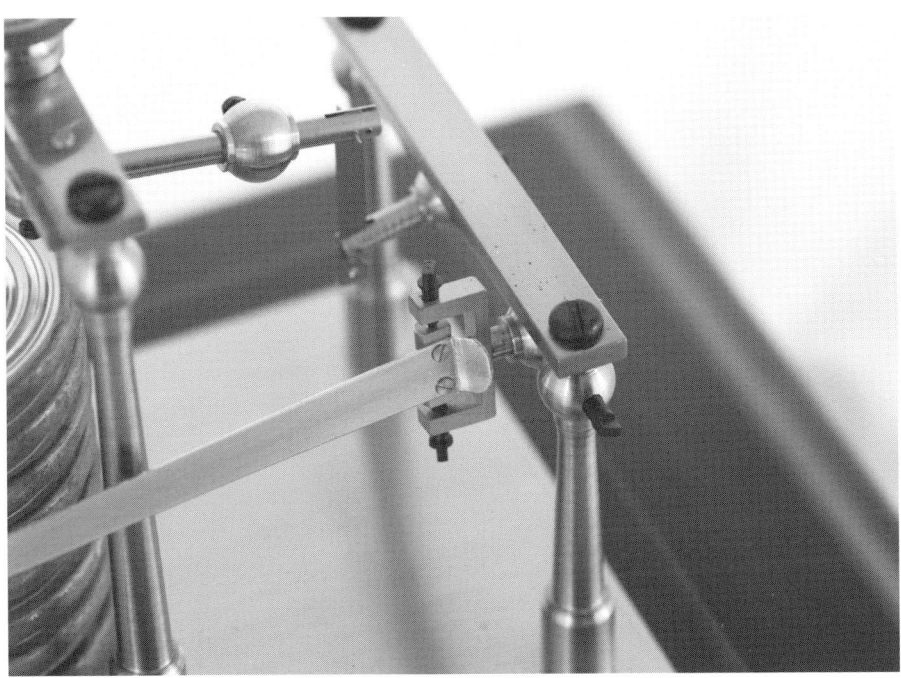

Fig. 3.9 Close-up view of a 'gate' arm used in the modern P. Collins of Merton barograph.

Fig. 3.10 Oak-cased barograph, c.1930.

Fig. 3.11 Top-down view of barograph in *Fig.* 3.10.

the drum and the bellows stack. It is fitted with a hinged gate-type arm. It is unusual in that these are nearly always found only in Met. Office fisheries-style barograph cases and not in a fully glazed case. This one is likely to have been made sometime in the 1930s.

The mahogany-cased flat-glazed barograph on four round pad feet shown in *Fig.* 3.12 is by Short & Mason, number 128847, measuring 14½ inches by 8½ inches by 7¼ inches high (37 x 22 x 18 cm). The mechanism, made of brass, is gold-plated as quite often found on Short & Mason items. The bellows are of single construction evacuated from the end, similar to those made by F. W. Darton & Co. in the 1960s and 1970s. I believe that the concertina unit used was the same as, or similar to, ones used on lorry radiator pressure valves. A few years ago, you could purchase this concertina-type tube in lengths: a simple tube with convolutions all along it. You could cut it to length and then seal the ends by soldering a cap on, allow for evacuation and thus create a single-cavity bellows. It was much easier to make than the highly technical manufacture of the multi-capsule bellows traditionally found.

One particularly interesting feature of this barograph is shown in *Fig.* 3.13 with the top and clock drum removed. This shows the Met. Office pattern clock. The brass plate that holds the bellows and the mechanism is also of the Met. Office type and would normally be found in the fisheries barograph design case, the bellows being relatively modern, and I would suspect that this can be reasonably accurately dated to around 1950. Short

Fig. 3.12 Mahogany-cased barograph by Short & Mason of 1950.

Fig. 3.13 Top-down view of barograph in *Fig.* 3.12 with drum removed.

& Mason were probably utilising components that were used for wartime production and general Met. Office use. As with *Fig.* 3.10, it is unusual to come across a Met. Office design clock on what is commonly considered a domestic barograph.

The fairly standard microbarograph shown in *Fig.* 3.14 (and in *Fig.* 3.15 with the top removed) is numbered 292/38, and therefore first calibrated in 1938, by Short & Mason. It is in a metal case with black crackle-finish paint and measures 14¼ inches by 8½ inches by 10½ inches high (36 x 22 x 27 cm). The barograph bellows are of the same design as the Meteorological Office fisheries barograph shown in *Fig.* 3.28 below. The arm, which is of the hinged gate type, has a throw-off device (*Fig.* 3.16), which is operated by a small lever just underneath the centre front edge of the case. The time-marker is provided by a long brass rod (*Fig.* 3.17), which is activated by a button outside, visible on the right-hand side of the case towards the base. The adjustment for altitude is above the bellows stack. The counter-balance is illustrated in *Fig.* 3.18 and can also be seen in *Fig.* 3.19 which shows the Short & Mason symbol on the brass bar. The metal case, which has two handles, left and right, to ease lifting and removing, has four angled pillars, which guide the case over and protect the drum and mechanism from damage when removing the large cover. The clock mechanism (*Fig.* 3.20) is of typical Met. Office pattern and is the same size as within the smaller drums, the drum on the microbarograph being 5½ inches (14 cm) in diameter and 7½ inches (19 cm) high. The charts are held on by spring-loaded brass clips

Fig. 3.14 Microbarograph by Short & Mason, 1938.

Fig. 3.15 Mechanism under the cover of the microbarograph in *Fig.* 3.14.

Fig. 3.16 Arm throw-off on microbarograph in *Fig.* 3.14.

Fig. 3.17 Time-marking device on microbarograph in *Fig.* 3.14.

Aneroid Barographs post-1902

Fig. 3.18 Counter-balance on microbarograph in *Fig.* 3.14.

Fig. 3.19 Detail of Short & Mason trademark on microbarograph in *Fig.* 3.14.

(*Fig.* 3.21) at the top and bottom of the drums.

These microbarographs have been much used by the Meteorological Office and provide more amplified indications of pressure changes. They were, however, found to have problems on board ship because of vibration from the engine or movement in high seas. The pens would often bounce up and down with this vibration or movement, thus producing a trace with thicker, more irregular lines than were usable, and therefore a cradle was designed to hold them on board ship to try to eliminate this problem. The cradle was quite basic, consisting of a board on which the barograph was fixed, with four strong rubber bungees attached above and below the instrument so that it was suspended in mid-air.

The model illustrated in *Fig.* 3.14 would have been for a land station or office use, but J. R. Bibby (1949) discusses the use of barographs on board ship and describes how the problem of vibration was overcome. Some ships may have been issued by this time with oil-dampened barographs. The oil-dampening method involves enclosing the whole bellows stack within a nearly sealed cylinder containing oil. To allow the expansion of the capsules, oil is forced through a small hole, which dampens the effect of sudden movement. This was first patented, number 641127, in 1949 by the Ministry of Defence. However, although I once heard about one of these oil-dampened barographs, I have never come across one and suspect that few were ever made and used.

A dark oak-cased thermobarograph is illustrated in *Fig.* 3.22, with bevel-edged glass, drawer and four simple pad feet; it measures 14½ inches by 8¾ inches by 8½ inches high (37 x 22 x 21 cm). The clock is patent number 3715.02 and is therefore of post-1902 design, probably dating to around 1910. Both arms are of the tension type; adjustment for altitude is above the bellows. *Fig.* 3.23 shows the mechanism of the barograph, which is fairly standard, with the additional thermograph mechanism. This is a coil of metal, which expands and contracts to record temperature. The glass panel near this coil has a hole (visible in *Fig.* 3.22), of approximately 1⅛ inches (2.8 cm) diameter, drilled into it to allow air in. This unfortunately often means that the mechanism gets dirty and corrodes more than it would with solid glass around it, but this hole is necessary for quicker temperature reaction on the sensor. There is a knurled adjusting screw just in front of the rear ink bottle, which is used for calibrating the temperature against the chart. The pens need to be able to clear each other, so that one can pass the other. This is usually solved, as can be seen in the illustration, by one arm having a longer nib than the other and being slightly set in front of it, so that only one time can be correct on these instruments. The charts are printed with both temperature and pressure divisions. Two ink bottles are supplied with dibbers: red ink for the thermometer pen and blue or purple for the pressure pen. Hygrometer (moisture) recorders use green ink.

Fig. 3.20 Clock under the drum of microbarograph in *Fig.* 3.14.

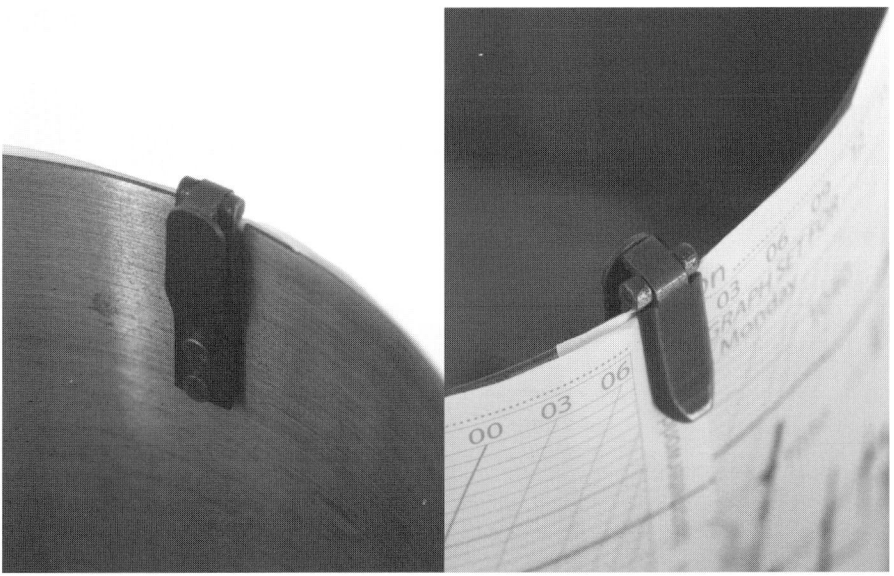

Fig. 3.21 Chart clip shown inside and outside drum on microbarograph in *Fig.* 3.14.

Fig. 3.22 Oak-cased thermobarograph, c.1910.

Fig. 3.23 Top-down view of thermobarograph in *Fig.* 3.22.

Another type of thermobarograph is illustrated in *Fig.* 3.24, measuring 16¼ inches by 9 inches by 11¾ inches high (41 x 22.5 x 29 cm). The clock drum is almost 7 inches (17.5 cm) tall and would use a chart that was marked in two halves: the top half for temperature and the lower half for pressure, unlike many thermobarographs that use the same chart to record both readings. The advantage of the tall chart is that both pens can be set to the correct time whereas in the previous example (*Fig.* 3.22) one pen has to be set at a different time or the pens will collide at some point. This one is marked R. & J. Beck, 68 Cornhill. Instead of a hole in the end glass, it has a brass mesh to allow air flow so that the temperature of the room is recorded more accurately.

Fig. 3.25 shows an attractive design of thermobarograph, dating from the 1930s, which looks less like a scientific instrument and more like a piece of furniture and would look good on a mantelpiece or desk of the same period. The combination of the clock and the two instruments makes it a well-balanced piece. It was made by Negretti & Zambra in an oak case with bevelled glass with a maximum height of 8⅞ inches by 24 inches wide by 5¼ inches deep (22.5 x 60 x 13 cm). The separate instruments of smaller proportion have drums 3 inches (7.5 cm) high and each has its own chart drawer beneath. The clock dial is silvered with black and red wax-filled engraving. The thermograph on the right uses the curved Bourdon style of temperature sensor connected to the pen rather than the more common spiral bi-metallic type. The barograph has bevelled glass panels to the front, top and side and the thermograph has two glass panels to the front and top and brass mesh on the side to allow air flow into the instrument.

The simple oak-cased barograph with round turned pad feet, flat glass, measuring 13¼ inches by 7½ inches by 6½ inches high (34 x 19 x 17 cm), attributed to Griffin & Tatlock Ltd, London, shown in *Fig.* 3.26, is numbered S6135, but with the S & M symbol on the brass bar (as in *Fig.* 3.19) connecting the two short pillars together. Therefore, it is plain that this was made by Short & Mason and not Griffin & Tatlock. It has a post-1902 design clock and has a gate arm fitted. The capsules, of which there are only four, have the narrow soldered edges as described in patent 198838 of 1922 (see Collins 1998, p. 207). This barograph is therefore plainly after 1922 and probably dates from the late 1920s or early 1930s. As there are only four capsules, they have been mounted on a brass pillar and raised up from the brass plate in order to make the item look more balanced.

Over the years, I have heard many people comment on the number of capsules a barograph has, believing that the greater the number of capsules, the better the instrument. As a rough guide, there could be a grain of truth in this, but on the whole it greatly depends on just how well the instrument is made. Certainly, the two-capsule model seen in *Fig*s 2.31 and 2.32 is often not such a good instrument over time, but whether a barograph has seven,

Fig. 3.24 Oak-cased thermobarograph, c.1905.

Fig. 3.25 Oak-cased clock, with thermograph and barograph by Negretti & Zambra, c.1930s.

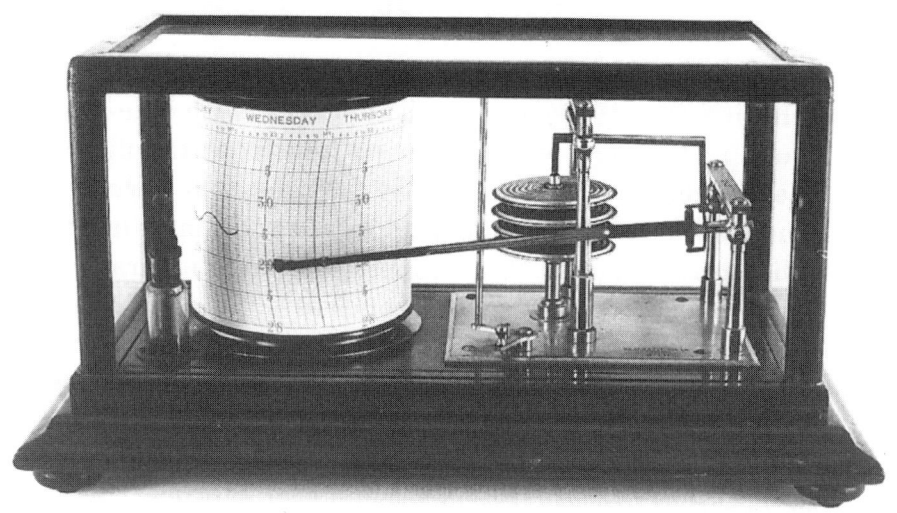

Fig. 3.26 Oak-cased barograph made by Short & Mason, c.1925.

eight or nine capsules usually makes little difference to its accuracy.

The sensitivity of a barograph and its accuracy are determined by the design of its levers and their ability to magnify and transmit the quite small up-and-down movement of the bellows to the arm. This I discovered when designing my own barograph. The bellows we had specially made to look like traditional models, screwed together like the old ones and with side nibs to each capsule, were made from beryllium copper and of a design that meant they reacted far more than the old ones ever did. Improved materials and manufacturing techniques over the decades have increased the amount of movement per capsule and mean that the pillars do not need to be as close to the bellows stack to amplify the movement. This has often meant that when we replace a bellows stack in a faulty old barograph we have to 'fix' three or four of the capsules together by drilling in the centre of them and screwing them together with a small machine screw so that they are fixed solid and do not move at all with pressure change. All they contribute is to make the bellows stack look original. There have been many types of barographs made, and some of the early French ones had a much larger stack of capsules. Negretti & Zambra in their precision barograph (see *Fig.* 6.13) used four stacks of capsules.

Fig. 3.27 shows what at first glance would seem to be a Victorian barograph. The case, measuring 15¼ inches by 9 inches by 9½ inches high (39 x 23 x 24 cm), has a typical Victorian appearance, with simplified gothic-style

Fig. 3.27 American walnut-cased Victorian-style barograph c.1905.

columns with incised carving. The case sits on small, simple bracket feet, making the whole item look like a Victorian museum cabinet. This particular model, although in original polish, is somewhat distressed, the polish having become crazed, chipped and damaged in various places. The handle is also of typical late-Victorian design. The movement has a tension arm adjustment for altitude above the bellows. The clock, however, has the patent 3715.02 and therefore is probably close to 1902 and not quite Victorian. There are comparatively few of this design of barograph. Sometimes they are far more profusely carved, but they always fetch an additional premium.

Fig. 3.28 illustrates a 1944 Met. Office pattern mahogany barograph with glass to front and left-hand side, with gate-type arm and altitude adjustment by the knurled nut above the bellows stack. The case measures 12½ inches by 6¼ inches by 8 inches high (32 x 16 x 20 cm). The bellows are particularly unusual on this instrument: they are a single unit instead of the earlier separate capsules. The whole concertina is evacuated and responds to air pressure as a whole as opposed to several capsules screwed together (see p. 63).

The throw-off arm for the pen goes through the case and can be used without opening the lid by the brass lever at the front. The button on the top of the case to the right of the carrying handle is a time-marker (see

Fig. 3.28 Met. Office pattern mahogany-cased barograph, 1944.

Air Ministry Meteorological Office, 1936). This button allows the observer to press the pen arm and cause it to record a time-mark; these should be about ⅛ inch (3 mm) long. If the marker is depressed gently, the pen will mark satisfactorily and the length of the time-mark can be kept within the desired limit. Time-marks could be made punctually to the minute at a fixed hour every day and any variation from this could be noted to the nearest minute and entered in the records alongside the appropriate time-marks on the chart before they were filed. This is mainly because it is difficult to keep the clock of any type of barograph completely accurate, and by marking the charts one has an accurate comparison of time, providing the chart is marked regularly with a clock of known accuracy. When the charts reached the Met. Office, they could record the data to the correct time rather than rely on the well-known variations of the barograph clock. I suspect that it also kept naval personnel busy making sure the barograph was recording correctly.

Fig. 3.29 shows the same instrument with the lid fully open. It is inscribed Short & Mason, who made a large number of these items for the

Met. Office. It has the letters M & O superimposed, a symbol of the Met. Office; the instrument is numbered 5882/44/58. This information tells us that it was made and first calibrated in 1944 and was recalibrated and probably overhauled in 1958. Many of these instruments were loaned to ships for recording air pressure and also some to fishing communities. The charts were then posted in supplied envelopes back to the Met. Office. Any inaccuracy in reading, i.e. setting for altitude, would then be advised to the users by the Met. Office and they would be instructed either to increase or decrease the reading by so many fractions of an inch. *Fig.* 3.29 also shows the Met. Office-style fixings for the charts: instead of one brass clip going down, these charts are held at the top and bottom by small brass sprung clips.

The clock is of unusual design as can be seen more clearly in *Fig.* 3.30. It is housed in a metal casing, which is screwed to the base of the barograph. The adjustment for time is underneath the cover, which is shown open in *Fig.* 3.30. The chart drum locates onto the central spindle and is held in position with a nut, which can be seen in *Fig.* 3.29 underneath the key. The Met. Office has always identified the clock as an independent instrument and it will quite often be numbered separately from the barograph. Often the clocks and drums are interchangeable, so if one fails then the clock can be exchanged for another.

The method of inking is also shown in *Fig.* 3.29. At the back of the right-hand side of the case is housed a glass bottle of ink and at the front on the right-hand side is a brass tube which holds a brass dibber for taking a drop of ink out and positioning it into the ink trough. There are two screw holes on the right-hand side of the case for fixing screws; although this method of fixing is not usual, it would stop the barograph from sliding around on board ship.

Great trouble was taken to find the best method of mounting these barographs on ships. A typical mahogany Met. Office fisheries barograph is shown in *Fig.* 3.31, measuring 12½ inches by 6¼ inches by 7¾ inches high (32 x 16 x 20 cm), by Wilson Warden & Co. Ltd, London, with the number 3348/40 engraved on the base plate along with the Met. Office symbol. The clock has its outer casing inscribed 'Met. Office W 1502/40', which is contemporary but numbered separately, as explained above. Unusually, the drum is also engraved with 'Met. Office S 1214/39', which therefore dates from 1939. It is not unusual for these items to be swapped around from one barograph to another as they were normally interchangeable, and this could have been done during refitting. There is little difference between this design of barograph and that featured in *Fig.* 3.28, except for the clock winding. *Fig.* 3.32 shows a large key fixed above the clock movement, which is exposed when the clock drum is removed. The more usual arrangement is illustrated in *Fig.* 3.29. A number of these Met. Office designs still survive:

Aneroid Barographs post-1902

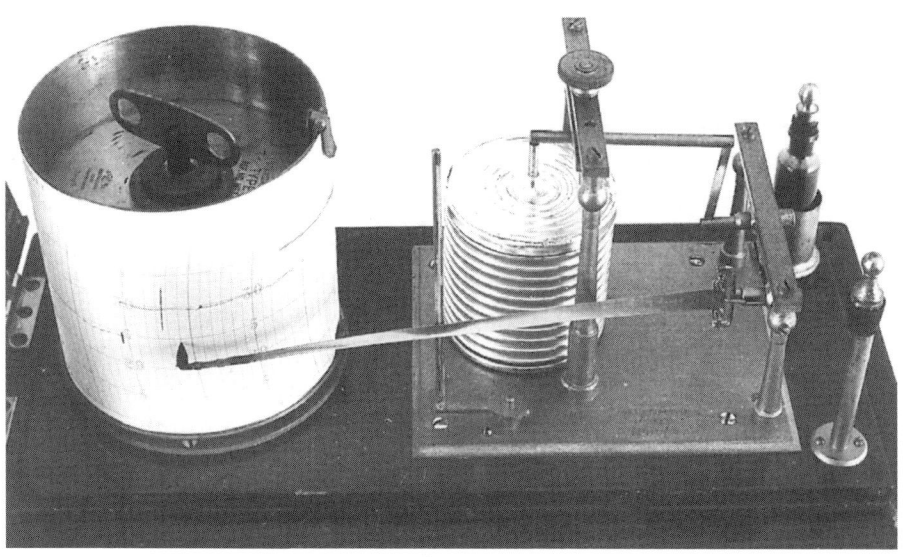

Fig. 3.29 Top-down view of barograph in *Fig.* 3.28.

Fig. 3.30 Clock unit from Met. Office barograph shown in *Fig.* 3.28.

Fig. 3.31 Mahogany Met. Office fisheries barograph, 1940.

Fig. 3.32 Mechanism of the fisheries barograph in *Fig.* 3.31 with drum removed.

although they seldom command as high a price as a fully glazed barograph, they are often preferred by nautical people. Although it has been well used, this model has maintained a very good colour to the case. It has probably travelled many thousands of miles on board ship.

Fig. 3.33 illustrates a barograph that is a rather odd mix of parts. Although at first sight it looks like a standard Met. Office fisheries barograph, the pillars of the movement (see *Fig.* 3.34) show the rather simpler design of later barographs. The clock is in a chrome-plated housing with chrome lid whereas the movement is brass. This is rather an odd mixture and, although it could have been changed, I suspect it is original to this piece. The bellows are the first I have seen of the Negretti & Zambra type used on a Met. Office fisheries barograph and, again, although it is possible they were changed, I somehow doubt it. It has an engraved plaque crediting it to Casella of London, and I suspect that this is quite a late barograph – perhaps 1960s or even early 1970s.

Fig. 3.35 illustrates an alternative type of Met. Office clock mounting which is in a taller cylinder but still screwed to the base plate of the barograph. This clock was manufactured by the Horseman Gear Co. Ltd of Bath, England, Met. Office Mark 2A, serial number 3347/60, which would indicate that it was made in 1960. *Fig.* 3.36 shows the drum cover which goes over the top of the drum clock. The cover, Met. Office standard drum type B, reference number MET2414, has a tensioned nut to hold this in position and the geared tooth seen at the top of the picture in *Fig.* 3.35 would locate into a gear inside the drum (not shown). The tension screw would then hold this in position on the teeth. The winding key would be inserted and usually left in position protruding above the drum.

Fig. 3.37 shows a mahogany-cased barograph with subsidiary dial, inscribed Munsey & Co. Ltd of Cambridge, who are almost certainly the retailers, as there is no reference to them being manufacturers. Below this is a registered number 428606; this model was almost certainly made by Short & Mason Ltd. The dial is a stylised dial without any divisions for inches of mercury but simply with weather interpretations linked to Admiral Fitzroy's words of 'Falls' for wet, warm or more wind and 'Rises' for cold, dry or less wind. The arm is of the sprung type, the clock mechanism of the 1902 patent. This design was registered in 1904 by Short & Mason and I believe this item to date from around 1920.

A popular type of barograph is the Short & Mason weather forecaster cyclo-stormograph shown in *Fig.* 3.38. The oak case with flat glass measures 14½ inches by 8¾ inches by 7¾ inches high (37 x 22 x 20 cm) and is mounted on four round pad feet. The circular base plate houses the hidden movement, referring to patent 22556 of 1904; it is inscribed Short & Mason *Tycos*, which is a trade name very often found on barometers and barographs by Short & Mason. This model is copyrighted 1921 and is probably amongst the earliest of this design. The pen arm is of the hinged gate style and, while we cannot be certain that this item is 1921, as it could have been a year or two later, it narrows down the period for the introduction of this

Fig. 3.33 A late Met. Office fisheries barograph, c.1960s.

Fig. 3.34 Mechanism of the Met. Office fisheries barograph in *Fig.* 3.33.

Fig. 3.35 Met. Office clock unit of 1960 by Horseman Gear Co. Ltd.

Fig. 3.36 Drum to go over clock in *Fig.* 3.35.

Fig. 3.37 Mahogany barograph with dial inscribed Munsey & Co. Ltd, c.1920.

type of arm to around this date. The adjustment for altitude can be seen in the foreground of the round base plate and inscribed set pen.

A similar Short & Mason cyclo-stormograph in an oak case is illustrated in *Fig.* 3.39. The mechanism is gold-plated. This instrument shows a number of interesting features. The clock is patent number 3715.02. *Fig.* 3.40 shows a view of the winding arrangement and escapement with the clock lid removed and has the clock number 8325. The ink bottle is held in a recessed hole in the corner of the front of the case. The mechanism itself is a single aneroid capsule below the slab, which incorporates patent number 22556 of 1904 by Short & Mason. According to the patent, the invention was 'to provide an instrument of improved construction which is capable of being fitted together and adjusted with considerably less trouble and expense than heretofore, and wherein the vacuum chamber is completely enclosed so as to be protected from injury.'

This seems to confirm that this type was designed to be a way of producing a cheaper barograph. None the less, this design usually works to the expected high standard of Short & Mason instruments (which can rarely be said of the double-capsule aneroid movement used in *Fig.* 2.31). The mechanism also has the later swinging gate style arm, and the plaque at the back, seen more clearly in *Fig.* 3.41, shows the cyclo-stormograph 'weather forecast' which is copyrighted 1930. This was designed to be used with

Fig. 3.38 Short & Mason cyclo-stormograph in oak case, c.1925.

Fig. 3.39 Oak-cased cyclo-stormograph barograph by Short & Mason, c.1935.

Fig. 3.40 View of the clock, with the lid removed, of cyclo-stormograph in *Fig.* 3.39.

Fig. 3.41 The plaque of the cyclo-stormograph in *Fig.* 3.39.

special barograph charts which had the letters A–J spaced alongside the inch measurements for the pressure (reprinted versions are available from www.barometerspareparts.co.uk). According to the reading in position on the chart, a forecast was given, either for a falling line or a rising line, between different letters of the alphabet. Despite the interesting arrangement of the movement – the gilding often surviving very well on these models – the case is a rather ordinary simple oak one with flat glass and with a simple plinth raised on four round wooden feet. On the circular base plate is inscribed Short & Mason, London, made for Army & Navy Stores Ltd, Westminster no. B63838.

Fig. 3.42 illustrates a small French mahogany-cased barograph with glazed front, the hinged cover held with two hooks and eyes, the case with machine-cut joints, made by Richard Frères, numbered 40686, and stamped 'MADE IN FRANCE'. It measures 7 inches by $4^{1}/_{8}$ inches by $5^{1}/_{4}$ inches high (18 x 10 x 13 cm), its exterior metal is nickel-plated, the movement is of traditional design and, as so often with French barographs, the adjustment for altitude correction is underneath by means of a small square key. *Fig.* 3.43, with cover raised, shows the clock which has a keyhole covered by a small brass turned knob engraved with a symbol of a double-ended key on the top of it. This helps keep the dust out but often these become lost. The advance and retard is underneath the sliding dust cover. The arm can be moved from the chart externally without the case needing to be opened. It would be so useful to have a definitive list of dates and numbers in order to date Richard Frères' barographs accurately. The 'Made in France' stamp suggests that this may date from the 1920s; certain British items began to be marked 'Made in Britain' or 'British Made' in the 1930s (see chapter 4, *Fig.* 4.4). Perhaps in time we will be able to trace some numbers to particular years, although many numbers on instruments can be misleading. I do tend to think, however, that the Richard Frères instruments were sequentially numbered.

Fig. 3.44 is a German Second World War barograph with a plaque on the top, engraved with swastika, eagle, large letter M and number 140 beneath, presumably instrument number 140. The mahogany hinged case is similar to British Met. Office patterns, although this particular model has glass both sides and to one end. The mechanism is of blacked brass; the capsules with flanges are soldered. There is no evidence of any side pipe from which to evacuate. Without ruining the item, it would be difficult to be certain how the vacuum is maintained: it could either be through the centre core, as with modern British barographs, or if each cell is individually sealed in a vacuum. Around the edge of each is a mark on the solder from where it may possibly have been evacuated.

The overall design is very similar to British barographs, with the altitude adjustment on the top bar between the two tall pillars; calibration is also

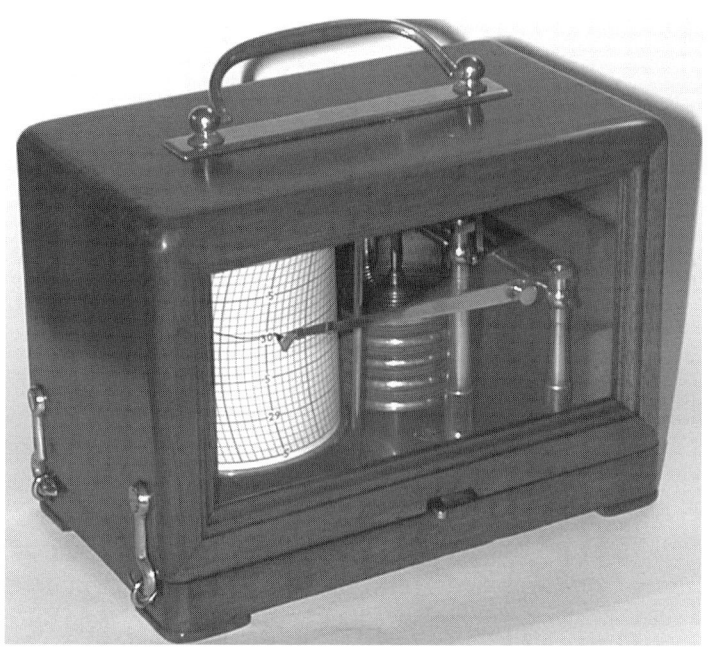

Fig. 3.42 Small French barograph by Richard Frères number 40686, c.1920.

Fig. 3.43 Mechanism under the lid of barograph in *Fig.* 3.42.

Fig. 3.44 German Second World War mahogany-cased barograph.

possible with a sliding linkage bar, and also on the arm pivot. Surprisingly, the pen arm is held on under tension, whereas British makers, by this time, were using arms that hinged under gravity, making less friction. It is possible that on board ship the hinged arm would swing away from the chart during motion, whereas the tension type arm would stay in its correct position. The most unusual device is the arrangement of the top of the bellows, which appears to have little practical use other than the screw at the top being adjustable to limit movement of the bellows downwards to restrict high pressure. Perhaps under warfare conditions this was thought necessary to stop damage to the bellows or the mechanism. Another feature is the use of plastic for the drum for the clock housing. Within the clock mechanism is a stag symbol which indicates that the clock was made by Tobias Bauerle and Sons, a clock factory in the Black Forest. After the Second World War the factory was dismantled, but started again at another site. The dimensions of this barograph are 11¼ inches by 6½ inches by 6½ inches (29 x 16.5 x 16.5 cm).

Fig. 3.45 is a rather plain oak-cased barograph, 13¼ inches by 8¾ inches by 6½ inches (34 x 22 x 16.5 cm), presented to C. D. Wilson, Esq. 'by the indoor and outdoor employees of Long Cross House' on the occasion of his marriage on 21 October 1925. The aluminium base plate is inscribed Negretti & Zambra R/1973. In 1921, Negretti & Zambra introduced a system of numbering their instruments: recording devices were prefixed by

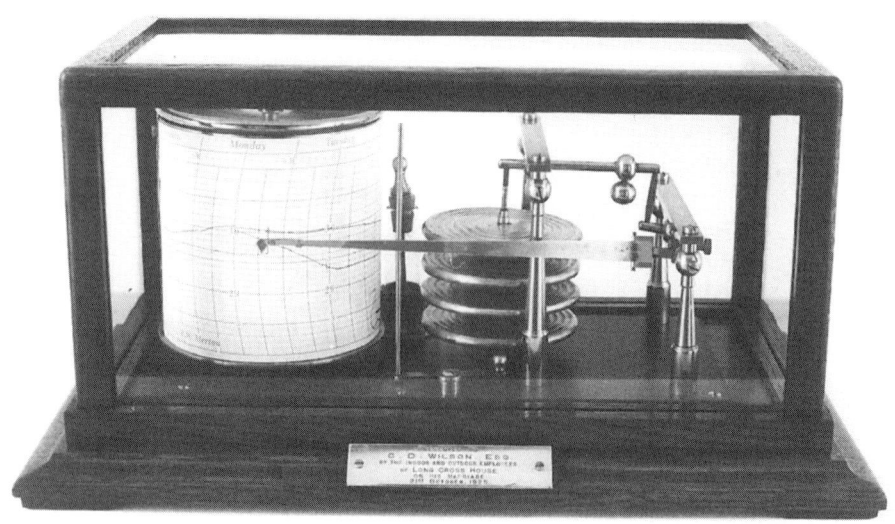

Fig. 3.45 Oak-cased barograph by Negretti & Zambra, 1925.

Fig. 3.46 Close-up view of 'adjustment' screw of barograph in *Fig.* 3.45.

'R' and meteorological instruments were prefixed by 'M', both starting with the number 101. As we can fix number R/1973, as given on this barograph, to around 1924–5, Negretti & Zambra must have made 1,872 recording instruments in these three or four years. I have another barograph inscribed R/1947 which can only be a short time before this model. Perhaps readers will have or will come across numbers which can be dated to help ascertain how many were made.

The simple case with flat glass has thin triangular flat feet with blackened brass base and lacquered brass movement, four larger than usual diameter capsules, measuring 2¾ inches (7 cm) in diameter, and additional counter-balance on the connecting arm. The clock uses the early designed gear of pre-1902, and the arm with the gate style fitting has a very unusual altitude-adjusting device, first patented, number 7500, in 1911. It is one of the few barographs with this patent adjuster that I have come across, seen more closely in *Fig.* 3.46. It is quite a complicated affair to make: the use is for large altitude adjustment and so one can assume that most were sent abroad to be used at higher altitudes than found in Britain.

Fig. 3.47 illustrates a mahogany-cased barograph by Negretti & Zambra, London, c.1905, the flat-glazed hinged lid held by a cranked hook to the simple wooden base. The brass work is nicely lacquered with a blued brass base plate. The throw-off arm lever is accessible only when the lid is opened and the bellows stack with four 2½ inch (6 cm) capsules is much larger than normal with soldered sides and evacuation spigots. Slightly later models of this style employed the 1911 patent for adjusting the pen arm (see *Fig.* 3.45). This model has a knurled adjusting screw on the end of the mechanism for finer re-calibration (see *Fig.* 3.48). By turning this screw, the length of the arm connecting to the rod up to the fulcrum lever is lengthened or shortened and thus a more accurate reading can be obtained, usually for setting in the factory during manufacture or re-calibration, not at home. This type of arrangement was common on good-quality aneroid barometer movements, and so it is not so surprising to find it used on the barograph mechanism but it is far less common.

From time to time keen engineers, usually retired, contact us to buy bellows so that they can make their own barographs. *Fig.* 3.49 appears to show one such instrument, possibly dating from about 1930. The bellows seem to be of usual design but the wooden case is surely not made by a professional cabinetmaker used to making barographs. It is very competently made but differs from the usual well-balanced case. The drawer section is over-sized and the appearance a little dumpy. The mechanism uses a normal clock and drum, but the pillars and levers are quite different from the traditional type. In the close-up view in *Fig.* 3.50, the arrangement of levers can be seen: the pillars are not tapered as they often are, there seems to be no counter-balance, and it all has a somewhat simple appearance.

Fig. 3.47 Mahogany-cased barograph by Negretti & Zambra, c.1905.

Fig. 3.48 End view of mechanism showing knurled calibration screw of barograph in *Fig.* 3.47.

Aneroid Barographs post-1902

Fig. 3.49 Amateur-built walnut-cased barograph, c.1930.

Fig. 3.50 Close-up view of amateur-built barograph mechanism in *Fig.* 3.49.

Casella sold many barographs, and certainly in later decades they bought them in from other makers. *Fig* 3.51 shows one such barograph, probably dating from the 1970s. It has a single concertina bellows (see *Fig* 3.52) as used by Dartons of Watford. The main pillars are simple in design but still retain some shape. An interesting variation (done for cheapness, I am sure) is the use of an aneroid arbour support to throw the arm away from the chart (see *Fig.* 3.53).

Throw-off arms vary over the years and with different makers. The simplest I have seen is probably by Gluck Engineering (see *Fig.* 2.6). Richard Frères used ones that operated from a lever outside the case and then moved it inside as can be seen in *Fig.* 3.54. Although this barograph was sold by Elliot Bros, it was clearly made by Richard Frères. *Fig.* 3.16 (above) shows the arrangement to throw the arm off used on a microbarograph made by Short & Mason for the Meteorological Office. A lever outside the case is pushed to the right and turns the rod which comes through the base plate on which is secured by a screw the throw-off arm and rod. Some also use a more elaborate design which we copied on our models (see *Fig* 3.55). It consists of an L-shaped brass piece with a small knob, fixed with a spring washer for tension and one end restricted by two short pillars, a more costly arrangement than some but one that makes the difference to some of us perfectionists.

Fig. 3.56 shows a modern mahogany barograph by Barigo of Germany. It measures $10^{3}/_{8}$ inches by $6^{3}/_{4}$ inches by $6^{3}/_{4}$ inches high (26 x 17 x 17 cm). The bellows are of the modern flange-soldered, single-evacuated capsule type, the clock is quartz driven and the pen arm holds a felt, ink-charged pen nib. The case hinges and has glass to five sides. In principle, modern barographs are the same as earlier ones. They are usually reliable and accurate, but somewhat lacking in their overall appeal when considered against the fine engineering of the late Victorian and Edwardian periods. Models such as these have been made for a number of years and are still made today. They are undoubtedly the antiques of the future. They are still mechanical, but it will probably not be many years before they are replaced by totally electronic models.

At first glance, one might think the ebonised cased barograph shown in *Fig.* 3.57 was a Victorian one, but closer examination shows the post-1920 gate arm design. The twin capsules are more modern, being flange-soldered, but could be replacements. The brass pillars are certainly of an early type of design, being nicely shaped and more individual than many standard barographs. I suspect it dates from the late 1920s or early 1930s. It measures $17^{3}/_{4}$ inches by $8^{5}/_{8}$ inches by $11^{3}/_{4}$ inches high (45 x 22 x 30 cm). The case looks a little oversized for the instrument when compared with the similar barograph in *Fig.* 3.58, as there is considerable space between the top of

Fig. 3.51 Barograph by Casella, c.1970s.

Fig. 3.52 Close-up view of the concertina-type bellows used in *Fig.* 3.51.

Aneroid Barographs post-1902

Fig. 3.53 Close-up view of the arm throw-off device used in *Fig.* 3.51.

Fig. 3.54 Close-up view of the arm throw-off lever on a Richard Frères barograph sold by Elliott Bros.

Fig. 3.55 Arm throw-off device used on our modern Merton barograph design.

Fig. 3.56 Contemporary German barograph by Barigo.

Fig. 3.57 Ebonised cased barograph with circular disc from the late 1920s or early 1930s.

the disc and the underside of the glazed case. Despite this, it looks original and makes an interesting variation to the standard barographs one sees. Note the single tall pillar near the disc and the quite different design of the linkages to move the arm.

There are some similarities with the oak-cased circular disc barograph shown in *Fig.* 3.58 but unfortunately neither of these two items is attributed to a maker. The twin capsules are of the more traditional side-soldered type. Like the barograph shown in *Fig.* 3.57, the movement has a single pillar supporting the linkage, but additionally this barograph has a semi-circular barometer dial. Perhaps the oddest feature of this barograph is that the recording disc is on the right of the movement. I am not sure I have seen any others with the recording part on the right. There is a registered number 479983 on the silvered dial which is not the numbering system used by Short & Mason, so perhaps this was a competitor trying a different design of dial. Judging by the small number of these large circular recording barographs that survive, they were not so successful.

Fig. 3.58 Large oak-cased barograph with circular disc and semi-circular dial, c.1920s.

4 Unusual Barographs and Altitude Recorders

This chapter will discuss some of the less common, unusual, rare, even one-off barographs, which are exciting to discover and a pleasure to handle. It is an intriguing business trying to discover the maker, designer, patents and so on for unusual barographs. I have been fortunate to have come across several of these unusual items, some of which are illustrated here, but there are doubtless more to be discovered.

I have noticed over the years that French instrument makers seem to have made a large range of unusual barographs. For this reason, I would recommend that the serious barograph enthusiast obtains a copy of *L'Histoire du Baromètre* by Bernard J. Maxant (2000), written in French but with illustrations of many varied instruments, including some very strange barographs, few of which will have survived. It is a useful reference book that contains information about some of the French makers, such as Eugene Bourdon and the Richard family, as well as the author's own family of instrument makers.

A barograph with an additional barometer dial makes it slightly more unusual and able to command a higher price. *Fig.* 4.1 illustrates a Negretti & Zambra barograph from the 1920s/1930s with subsidiary dial with jewelled movement, an eight-capsule stack and hanging thermometer in a hinged oak case with a drawer. This type of barograph is very popular and usually fetches a premium price because of the additional dial. Operationally, it can have a tendency to lag, as there is more call on the energy provided by the capsules to drive the barometer hand as well as the pen arm. While this is often very slight, this model by Negretti & Zambra has jewelled mounts to the four main pivots, which help to reduce the friction to a minimum. In practice, it is debatable whether the jewels are really of any benefit as the normal pointed bearing commonly used does seem to work quite adequately. Doubtless a jewelled bearing might help, but the reduction in friction must indeed be a very small gain.

The dial (which can be seen more clearly in *Fig.* 4.2) is probably acid-etched rather than hand-engraved. The fusee chain and linkage connecting the arbour of the hand to the movement of the arm can also be seen at an angle of approximately 45 degrees pointing towards the 31-inch position on the dial. The thermometer scale is etched on the glass in Fahrenheit and centigrade and is held at the top of a triangular rod clamped in with

Fig. 4.1 Oak-cased Negretti & Zambra dial barograph, c.1920s/1930s.

Fig. 4.2 Close-up view of dial of barograph in *Fig.* 4.1.

Fig. 4.3 Top-down view of barograph in *Fig.* 4.1.

a brass screw at the top. The barograph has a traditional eight-day clock mechanism and drum of the pre-1902 design; the arm is of the gate type. The pen setting knob is located behind the base of the open dial on the metal base plate and the ink bottle is on the opposite side of the base plate to the thermometer. The top view, shown in *Fig.* 4.3, illustrates just how congested the whole mechanism is. The lid is hinged and prevented from opening fully by a folding brass arm visible towards the top of the picture. It is without doubt an advantage to have a hinged lid; not all hinged lids have stops but most do, which reduces the damage possible when the lid is lifted or replaced carelessly.

Fig. 4.4 is an interesting mahogany-cased barograph with subsidiary dial, bevelled glass all round, with plaque attributing it to John Trotter, 40 Gordon Street, Glasgow. The base plate is engraved 'British Made' on the front left-hand side. The mechanism is of polished aluminium with an early-style clock mounting with tension arm for the pen and side evacuated capsules, but with flattened edges, which are after the 1922 patent. Underneath the drawer, written in pencil, is the date 1937, which is possibly the manufacturing date, although this is not certain. The case measures 14½ inches by 9 inches by 9 inches (37 x 23 x 23 cm) and was originally ebonised as is often found with polished aluminium movements. It has been stripped and re-polished some considerable time ago and is a reasonable colour mahogany, as preferred nowadays. Traces of black polish still survive around the inside of the drawer, and the inside of the case is still ebonised, obviously being harder to strip and polish without removing all the glass.

Fig. 4.5 is a mahogany dial barograph after the Short & Mason design,

Fig. 4.4 Mahogany-cased barograph with dial by John Trotter, 1937.

the clock being of pre-1902 style, and the pen arm is of the tension type because of the congestion of the levers. The knurled tension adjuster for the arm is behind the arm this time and is not accessible from the front. The altitude adjustment is by a knurled knob operating through the base plate. The dial is mounted with a curved mercury-filled thermometer, divided for centigrade and Fahrenheit. If compared with other dial barographs, it will be seen that this type has a hair spring and spindle assembly mounting plate, which is clearly visible in the middle of the open dial, running vertically, not, as is more usual, horizontally (see *Fig.* 3.5 in chapter 3). It has quite heavy block feet and dates probably from around 1920.

Fig. 4.6 shows an early twentieth-century barograph, possibly dating to around 1910. The case is rather nice with its rounded corners; the small bracket-type feet are also rounded to follow the curve of the case. All this means extra work when making and is usually a sign of higher than normal quality. This barograph has gold-plated brass work to the mechanism and a blued brass base plate and clock drum. It is easy at first glance to think that the blued brass is steel, but it is indeed brass that has been treated. I believe that this was a method to avoid the brass corroding. It was no doubt cheaper than gold plating, but I always feel that the dark blue finish of the base plate is rather dull and less interesting than gold-plated or lacquered brass. These base plates are generally in good order but may need a light clean

Fig. 4.5 A mahogany-cased dial barograph, c.1920.

over with thinners and sometimes a clear lacquer to maintain a suitable matt shine. The glass on this instrument is very thick bevelled glass which makes the glazed lid much heavier to remove.

Fig. 4.7 is a very similar barograph, but in a very simple walnut case, probably a little later than the barograph in *Fig.* 4.6 but not much. It is most likely by the same manufacturer which might be Short & Mason as they gold-plated their brass movements more often than any other maker. Note the very large ink bottle: this one has been replaced, but these large ones were often made of cut glass and are rather nice.

Barographs with dials have always been popular selling items, the extra dial making them more appealing. *Fig.* 4.8 shows a late barograph sold by Negretti & Zambra with a semi-circular dial in a simple oak case. This one dates from the 1960s and has the standard Negretti & Zambra style bellows. The main difference with this type of barograph is the method by which the movement of pressure is transferred to the hand of the barometer dial. A fine chain linked to the pen arm spindle on the right enters the brass drum seen behind the centre of the barometer hand. Inside there, it is wound on a small brass V-shaped pulley wheel which is also mounted with a hairspring, much like an aneroid barometer movement, so as to keep some small amount of tension on the hand. It is very fiddly to dismantle and re-assemble but does seem to operate quite well.

Fig. 4.6 Early twentieth-century barograph.

Fig. 4.7 Early twentieth-century barograph in simple walnut case.

Fig. 4.8 Semi-circular dial barograph by Negretti & Zambra, c.1960s.

Fig. 4.9 is an example of the patent number 7323 dated 1891 (see patents list in the Appendix); the names on the patent are Georges Meyer and Antoine Rédier of Paris. Antoine Rédier (1817–1892) is a famous French instrument maker who continued a very successful business, making many types of instruments including numerous aneroids. Many survive with a very distinctive logo designed around the four letters 'ARED' where the 'R' is reversed against the 'E', the 'D' printed over the 'E' and the 'A' at 90 degrees and larger over the other three letters which are all entwined. This barograph has a mahogany glazed case, of dimensions 6 inches by 4¾ inches by 10¼ inches high (15 x 12 x 26 cm), with the front hingeing down. The front cover also has mounted within it an enamelled-face clock which is a very rare feature.

I have handled another model, number 36, and inscribed 'J. Hicks', who obviously sold the item. The main reason for this design was to overcome the friction of the pen on the charts. As the pen is not in contact but moves freely until a hammer device taps the pen to make a mark on the chart, there is no friction on the recording side of this barograph. Few of these designs survive however. The drum is mounted horizontally and is designed for charts to run over the front bar, which is held away on a brass tube supported by two sprung wires. This provides tension for the chart. The clock raises a rather complex-looking frame which it drops a few times an hour to knock the pen arm and make the small circular nib deposit a

Fig. 4.9 A rare vertical French mahogany-cased clock and barograph, c.1891.

small dot of ink on the chart. Invariably, these pen arms are missing on the instruments. To replace them, we have used a narrow piece of spring steel (recycled from a clock spring) with a suitable hole in one end to fit the circular nib; the fixed end just slides under a strip of steel. It is important to get the length correct, as well as the tension, so the frame that drops onto it at regular intervals pushes the arm down and allows the arm to come back up when the frame lifts up again as the clock rotates.

This design does seem to be rather complex and it is probably no wonder that they did not succeed in becoming the mainstream type of barograph. The engineering is very high quality and the ones I have handled have mostly had the brass parts gilded which prevents the normal corrosion that happens after many years. The clock also revolves the chart continuously but the charts are designed to be changed weekly as the time on the chart does not continue over the join in the paper as with the usual type of barograph. Above the drum is a scale divided into inches with weather words etched or stamped. For sales in other countries, a different scale would obviously be fitted.

The model shown in *Fig.* 4.9 is numbered 151 and probably dates from soon after the patent of 1891. The single large capsule, 3⅜ inches (8.5 cm) in diameter, and ¾ inch (20 mm) thick, will have internal springs within it, like the Vidi patent. The case is similar in appearance to Richard Frères' cases with machine-cut joints on the corners. The movement has a nineteenth-century French appearance with the gold plating. On the right-hand side is a small bolt which, when opened, allows the barometer movement to hinge upwards on the small top frame (this is hidden underneath the word 'Very Dry' in *Fig.* 4.10), possibly to allow changing of the charts and access to the pen arm. The adjustment for altitude is by a small knurled adjuster above the top of the large capsule, which can be seen in *Fig.* 4.10. Maxant (2000) has more detailed information on this type of wall barograph by Antoine Rédier.

Although they are not very common to find, there was a variety of these early barographs made. *Fig.* 4.11 shows a very decorative one with an enamel clock mounted above the recording drum and housed in a gilded brass case. It measures 7 inches by 5⅜ inches by 11¼ inches high (18 x 13.5 x 28.5 cm); the front bevelled glass door is hinged for access. The scale is marked in inches of pressure.

Patent number 20800, dated 1911, illustrates an improvement in recording mercury barometers by Giuseppe Agolini of Italy, a rather different type of recording barograph. The detailed drawing (*Fig.* 4.12) shows a rod passing through the column of mercury with a large float resting inside the vacuum and lifting a recording pen up on the clock drum nearer the base. One of Agolini's curious barographs was displayed in Florence in 2013 among some eighty other instruments in the exhibition *Dal Cielo alla Terra*. This barograph was not quite in accordance with the drawing in *Fig.* 4.12, but was constructed with a metal frame rather than the wooden one of the patent. It is common to find variations between patent and actual instruments as improvements were continually made on the original design. Anita McConnell (2013), reviewing the exhibition in issue 1 of the *Royal Meteorological Society History Group Newsletter*, introduces this very unusual type of barograph:

> My own favourite is the barometrograph by Giuseppe Agolini of Parma, who in 1912 constructed an elaborate recording mercury barometer and bimetallic thermometer whose design

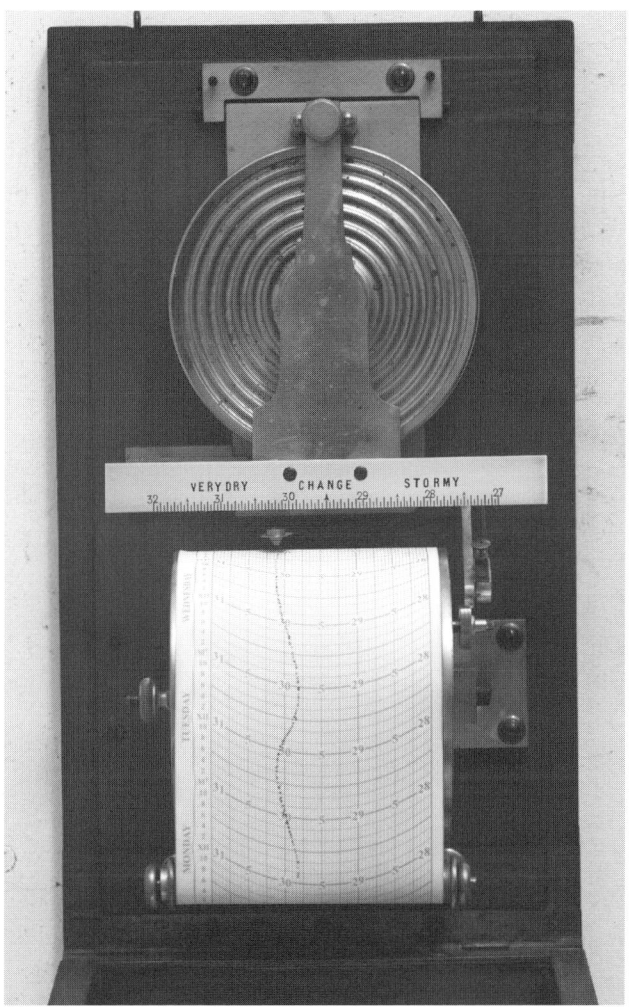

Fig. 4.10 Vertical barograph in *Fig.* 4.9 with lid open.

had eluded Christopher Wren and Robert Hooke. After satisfactory testing at Bera (Milan) Observatory, Agolini began producing instruments for other Italian observatories, but this production was cut short by the First World War when the factory switched to altimeters and other aircraft instruments. The description below is taken from the *Catalogue of the Bera Astronomical Museum*.

'The mercury barometer holds 18 kg of mercury, the column being immersed in a base comprising two parts, the upper of which can be isolated from the main tank. The base rests on three legs, with levelling screws. The tube has a cylinder at

Fig. 4.11 Gilt brass cased clock and barograph by Antoine Rédier, c.1875.

its upper end within which the free surface of the mercury column is located. A disc float resting on this transmits the height variations of the mercury column to the writing instruments by means of a rigid structure. A rod is fixed to the float and drops into the mercury column as far as the lower structure. Here the end of the rod connects with a cross to which three thin vertical mounts are attached outside the barometric tube. The upper part of these mounts emerges from outside the mercury and the arm of the lower writing needle is attached to them. The mount complex can only move vertically, as a collar is fixed to its upper part, fitted with small rollers. This rigid

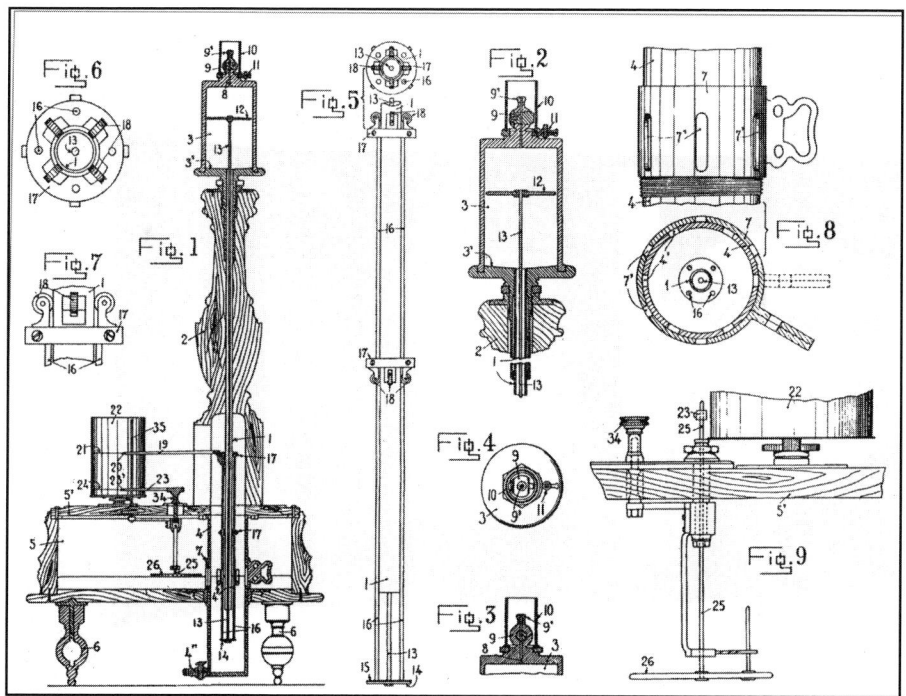

Fig. 4.12 Patent mercury barograph from 1911 by Giuseppe Agolini of Parma.

group is designed to avoid introducing resistances or errors in the indication of the mercury level due to thermal expansion. Temperatures are measured by means of a bi-metal spring connected to the upper writing needle. The two needles write on paper mounted on a rotating cylinder driven by clockwork, to revolve once a week.'

Fig. 4.13 illustrates a barograph labelled 'Jules Richard', who took over the brothers' business but still traded under the old name (see Brenni 1996b). It is very unusual and few readers will be likely to see such an example, although I have heard that some exist. I am uncertain of this instrument's date or if many were made. A French catalogue of the 'Establishment of Jules Richard' survives, dated 1931, also entitled the 'old house of Richard Frères'. Within this are two illustrations of this style, one almost identical to the model shown in *Fig.* 4.13 and another larger one, which has a longer weight or series of weights in a case underneath with a wooden door, the more usual type of barograph arrangement being visible through a glazed cabinet situated on top. The whole design is similar to other barographs except that the capsules are not sprung internally. Instead, a weight is attached

Fig. 4.13 A weighted barograph by Jules Richard, possibly 1920s.

to the capsules to balance the weight of the atmosphere exactly, thereby allowing any variation in pressure to expand or contract the height of the capsules and indicate this on the clockwork-driven chart drum.

The attractive dial barograph illustrated in *Fig.* 4.14 is the first I had seen of this type. It is made by Short & Mason and incorporates their 1932 copyrighted dial, which is more usually found on round aneroid barometers or the hexagonal oak-cased aneroids of the period. The earlier wall aneroids also had a little window in the dial behind which a small type of flag showed blue or red according to whether the barometer was rising or falling. I have not seen this flag indicator system on an extant barograph, and it is likely that the cost or complications over the extra friction created on the mechanism meant that the design was abandoned on later models, certainly after 1945. The capsules are also an indicator of later age: *Fig.* 4.14 shows the concertina-type bellows commonly used by F. W. Darton and Co. in the 1970s. The case is of oak and is of the standard dimensions.

Fig. 4.15 is a small, brass-framed carriage clock-style barograph, c.1885, measuring 5⅜ inches by 2¼ inches by 5 inches high (13.5 x 6 x 13 cm), with three capsules, the clock winding from underneath. The case is held down

Unusual Barographs and Altitude Recorders

Fig. 4.14 An oak-cased dial 'stormoguide' barograph by Short & Mason, c.1930s.

Fig. 4.15 Small brass-cased barograph, c.1885.

by two sliding brass bars underneath the base plate to left and right. There is a retailer's plate 'F. Kuhn, Optiker, Lucerne', indicating that it was sold in Switzerland but I cannot be certain of its origins: there are no manufacturer's marks on any of the metalwork. Underneath is a small knob to adjust for altitude. The unusual assembly magnifying the movement of the capsules, very similar to French aneroid barometers, extends the lever that is counter-balanced to a short lever attached to the arm. This has a small screw to adjust the amount of magnification by sliding the link up or down this small arm. The arm itself, which can be seen in *Fig.* 4.16, is shaped to go round the pillar and has the usual tensioning knob to hold the arm against the paper. This illustration also shows the clock movement without the drum. Clock experts may be more accurate in dating this from the style of the case, the carriage box and the clock movement, but it appears to be late Victorian. An unusual feature is the nib, which is slotted on to the end of the arm and held in position with two screws. A similar design is incorporated for the bar to hold down the charts.

Charts for these instruments are very difficult to come by: original ones seldom survive to give us an idea of the design for uncommon barographs. To design a chart, you need to place paper around the drum and run the clock for a given length of time, marking the chart when it starts and finishes, and not allowing the chart to go over the join. In practice, one can hardly run it for exactly seven days, but the nearer to seven days the better, then calculate what seven days would be and you get the length of the time divisions needed per week and per day. To calculate the pressure range, you need to see what range it measures on a blank piece of paper over a period of time which gives enough of a range that you can calculate what divisions are needed on the chart. This is all a very time-consuming process. Once you have the correct measurements, then a graph can be drawn on a computer and the artwork produced to print charts. However, getting the right paper is another problem, which is discussed in chapter 6.

This attractive little barograph is complete with its carrying case, of typical French design with thin leather (or paper?) covering, hinged lid and visible opening, as seen in *Fig.* 4.17, with its front panel removed. This is the first of this type that I have come across and it has to be said that the small size makes it difficult to remove the case without fouling the arm. This is hardly a carriage clock, but I can envisage interested observers taking this on a journey with them and using it, perhaps in their bedroom, for a period of weeks or months whilst they are away. It is certainly a very interesting item, which, after we had had it in our workshop, was destined for use on board a boat in a specially made gimbal. The carrying case is in remarkable condition for its age. The only thing apparently missing is its handle, which one assumes would have been leather and has thus become detached.

Fig. 4.18 illustrates a two-week recording barograph by Negretti & Zam-

Fig. 4.16 Top view of barograph in *Fig.* 4.15.

Fig. 4.17 Brass-cased barograph in *Fig.* 4.15 shown in its carrying case.

114   *Unusual Barographs and Altitude Recorders*

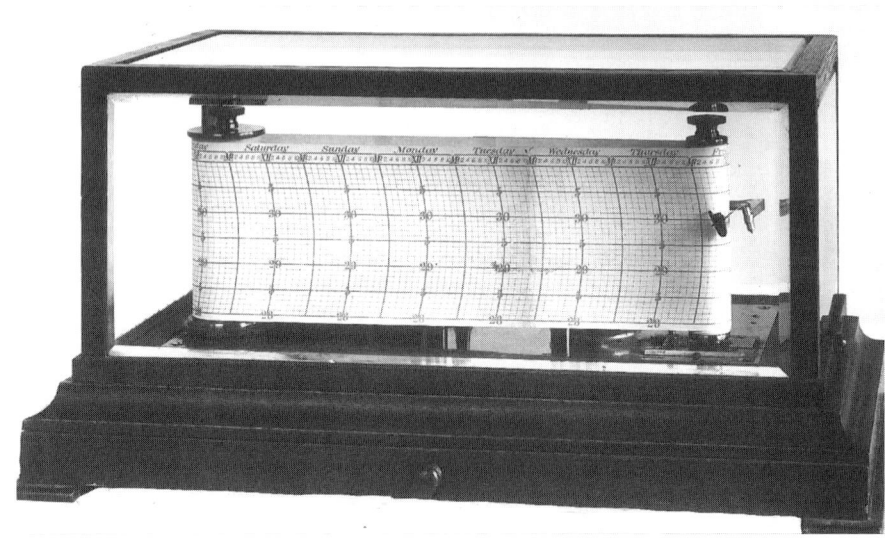

Fig. 4.18 Oak-cased two-week recording barograph by Negretti & Zambra, c.1920.

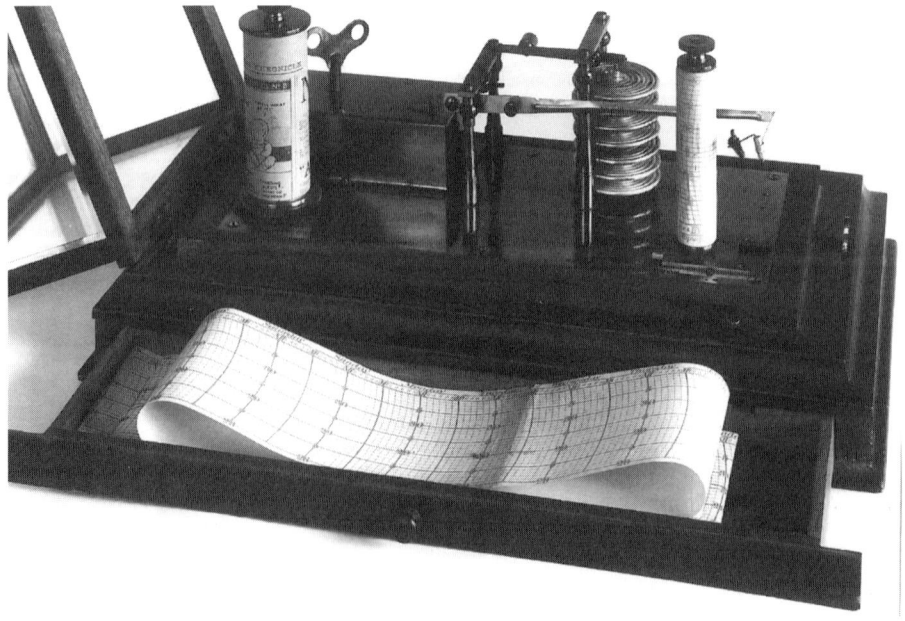

Fig. 4.19 View of barograph in *Fig.* 4.18 with lid open and chart removed.

bra. The oak case is fitted with five bevel-edged glass panels and is hinged, as can be seen in *Fig.* 4.19 with the chart removed and the lid open. The clockwork is housed in a rectangular box behind the larger roller that drives the roller on the lower gear. Tension of the chart is achieved by adjustment of the smaller roller. The arm is of the self-sprung type; the bellows are of the pre-1922 design with soldered sides and spigots. The basic barograph mechanism is the same as the majority of barographs of the standard type. The unusual feature is the pen, which comes around the chart and has been fitted with a capillary nib, as opposed to the traditional V-shaped bucket nib. The arm has been shaped to strengthen it, which could be required because of the extra weight of the pen. This type of pen is not usually found until the 1920s, and for this reason I would consider that the barograph may be as late as 1920. The rollers have paper stuck around them, which is a later addition. I suspect the rollers were originally rubber but in time have perished, so paper has probably been used to tidy up the rubber and make a more even surface for the charts to run on. Notice that the pillars are still the same design as Richard Frères' barographs.

The drawing in *Fig.* 4.20 is from page 9 of L. Casella's 1894 *Illustrated and Descriptive Catalogue of Recording Instruments by Richard Frères* and shows a remarkably accurate barograph called a 'Statoscope' or 'Extra Sensitive Barometer'. If raised 3 feet (90 cm), a movement of 1/10th of an inch (25 mm) shows on the pen. The picture is a little misleading as there is a tap shown which is normally evidence that the instrument was for remote pressure readings – normally water pressure to calculate the height of water in a mine or wherever it was needed. I suspect it could have been made as a more normal barograph as well. The catalogue also mentions a 'Brontometer' constructed for them by Mr G. J. Symons, FRS, but continues to discuss the statoscope which makes it uncertain whether the 'Brontometer' is actually a different design or the same as illustrated. I have not seen any of these designs but do not doubt that some may exist.

My first introduction to an unusual barograph was a pocket altimeter. *Fig.* 4.21 is a similar French recording altimeter. This one is in its original wooden case, with provision for charts, key and ink bottle (see *Fig.* 4.22). These are fascinating small barographs which are rare, and this was the first one we found which was in its original boxed case. The hinges and fitments are nickel-plated. *Fig.* 4.23 shows a close-up view of the clock and barometer movement with its protective cover removed. It measures 4¾ inches by 3½ inches by 1⅜ inches high (12 x 9 x 3.5 cm); the box is 7¼ inches by 5¾ inches by 2¼ inches high (18 x 14.5 x 6 cm). The key, which is shown in *Fig.* 4.24, is like the Richard Frères' barograph keys, but one end is squared to locate into the chart rollers to turn the chart roller along. The small hollow square is for the altitude correction adjustment to the barograph and the other end of the key is for winding the clock.

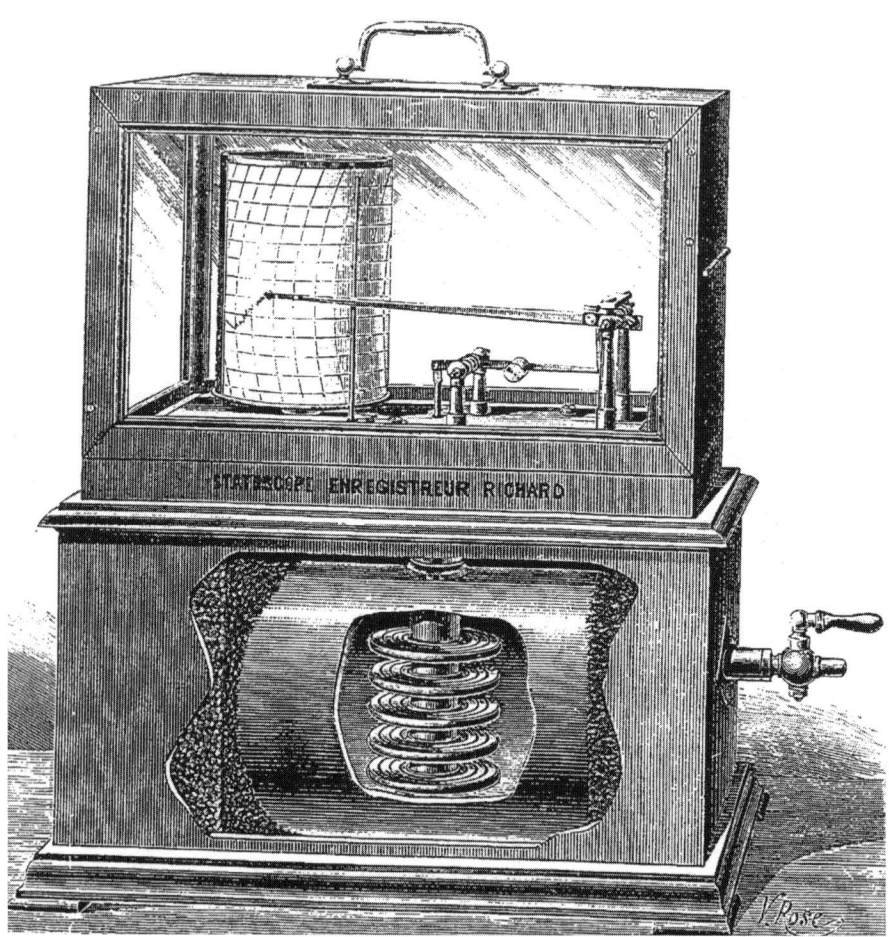

Fig. 4.20 Statoscope or 'Extra Sensitive Barograph' by Richard Frères illustrated in Casella's catalogue of 1894.

I have been unable to find out the makers, but the dust cover has the initials 'SGDG' inscribed into it, the society to which Richard Frères belonged, although I am not certain that it was actually made by Richard. Below this is another pair of initials, possibly back to back, which may be 'JL'; it is also inscribed 'Optique Santi Marseille' who were possibly the retailers. According to Edwin Banfield in *Barometers: Aneroid and Barographs* (1996, p. 136), the firm T. Cooke & Sons Ltd of London were advertising these in a 1913 catalogue as being available in three types for use up to altitudes of 4,000, 8,000 and 16,000 feet. The charts with this model are calibrated up to nearly 2,500 metres. It was clearly not a popular instrument as few are seen in auctions which makes those that are available of greater interest to collectors and so often more expensive items.

Fig. 4.21 Small French altitude recorder within original box, c.1910.

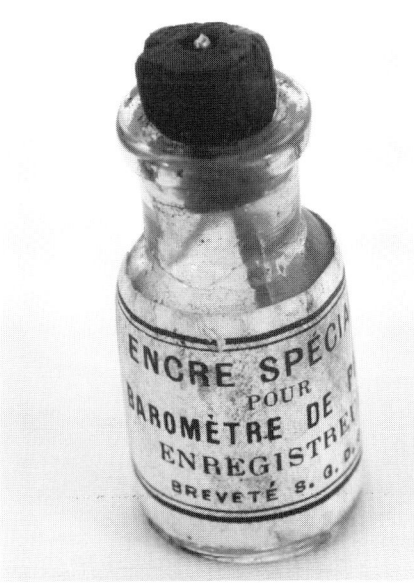

Fig. 4.22 Original glass ink bottle, complete with original paper label, supplied with the altitude recorder in *Fig.* 4.21.

Another type of altitude recorder is shown in *Fig.* 4.25 in a mahogany case with metal bandings by Short & Mason of London. It measures 9 inches by 4½ inches by 5½ inches high (23 x 11 x 14 cm). This model is Mark IB, minus 1,000 feet to plus 20,000 feet. The bellows are a single evacuated stack; the base plate and pillars are of lightweight aluminium. This rather sad example has the pen arm mechanism missing and also the clock. It could date from as late as 1970 but is probably typical of items from the 1960s. These designs seldom changed very much. *Fig.* 4.26 shows the mechanism withdrawn from its case.

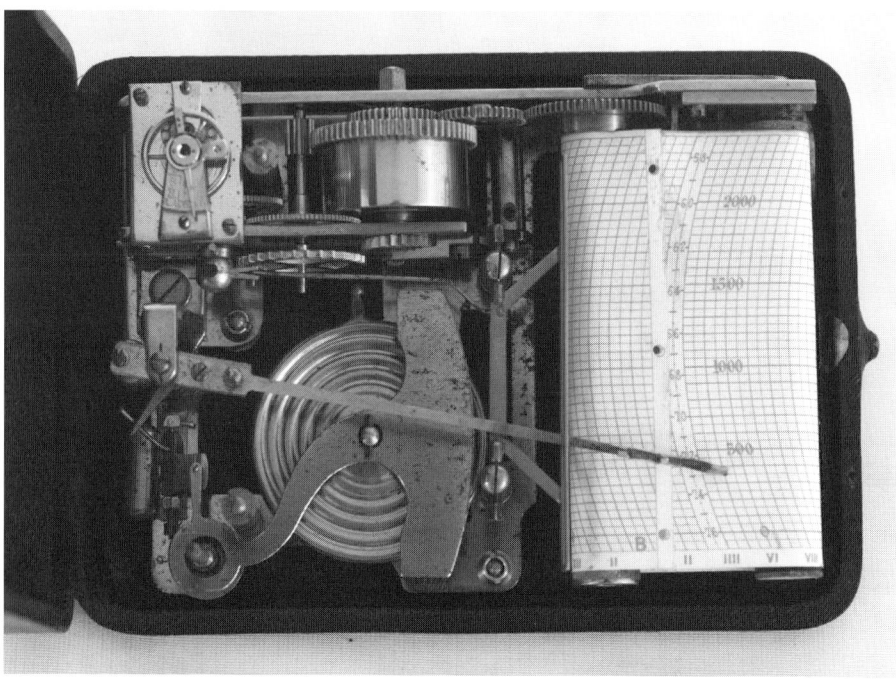

Fig. 4.23 Altitude recorder in *Fig.* 4.21 with lid open and dust cover removed to display mechanism.

In *Fig.* 4.27 can be seen a rather battered and derelict altitude recorder for airplanes, balloons or gliders. This one is of German origin, still with an original chart. In the top right-hand foreground of the picture can be seen a lead seal, which presumably would stop the item from being tampered with during use. *Fig.* 4.28 shows the interior of the recorder: the whole mechanism withdraws under the wooden case for winding and changing of the chart. The case is probably of beech and measures 7¾ inches by 4½ inches by 5½ inches high (19.5 x 11 x 14 cm). As these would not be required to last for more than a flight, the clocks on

Fig. 4.24 Double-ended key for use with altitude recorder shown in *Fig.* 4.21.

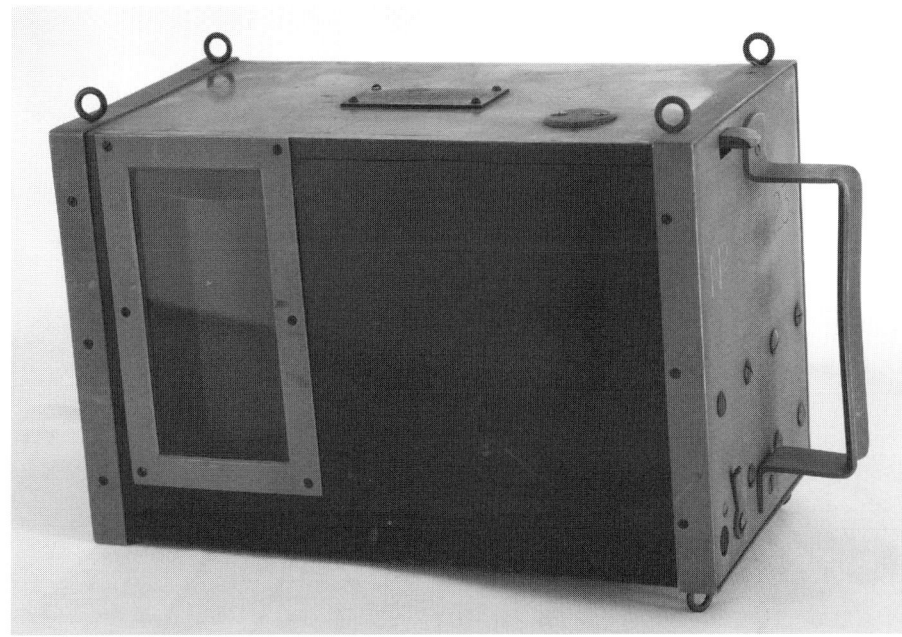

Fig. 4.25 Altitude recorder by Short & Mason, c.1970.

Fig. 4.26 The inside of altitude recorder shown in *Fig.* 4.25 (parts missing).

Fig. 4.27 German altitude recorder in wooden box (rough), c.1910.

Fig. 4.28 The inside of altitude recorder shown in *Fig.* 4.27 (parts missing).

Unusual Barographs and Altitude Recorders 121

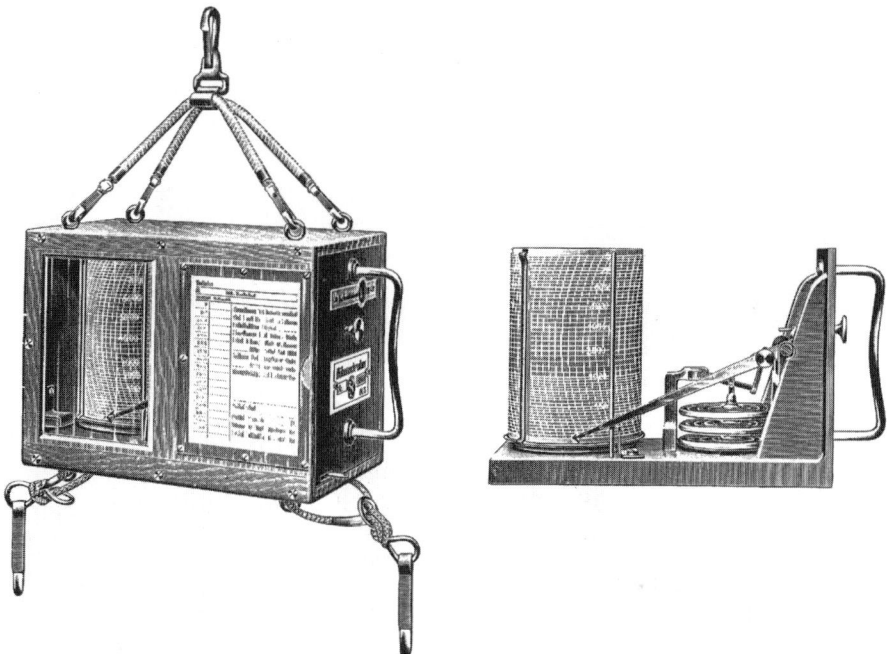

Fig. 4.29 Altitude recorder from Emil Schultz catalogue of 1910.

these instruments are usually only 30 hours. This one quite likely originated from Emil Schultz of Hamburg. It certainly shows great similarity to the one illustrated in *Fig*. 4.29 from the Emil Schultz catalogue of c.1910. The catalogue illustration shows how it could be secured to reduce the effect of vibration on the recording, quite similar to the mounting of Meteorological Office barographs on some weather ships.

Fig. 4.30 shows a barograph with a circular chart instead of the normal strip chart. The case is of nickel-plated metal with bevel-edged glass fitted, sitting on an ebonised wooden base, measuring 13¾ inches by 7¼ inches by 9 inches high (35 x 18 x 23 cm). All the interior metalwork is similarly nickel-plated. The clock, as can be seen in the rear view with the cover removed in *Fig*. 4.31, is mounted horizontally facing forward on a large metal bracket. The circular chart therefore gives a continuous recording if not changed, although still divided for seven days. This type of disc-recording chart was first patented, number 1900, in 1894.

The barograph also has an aneroid dial with mercury-filled curved thermometer. The dial is chemically etched and the definition of the lines is not good; the whole dial is nickel-plated and the characters are not black filled as in traditional silver dials. The arm is of the hinged gate type and therefore can surely be dated after 1920. Edwin Banfield (1996, p. 133) dates

Fig. 4.30 Late 1920s' nickel-plated cased barograph with circular chart.

this to c.1925 and I certainly think that the later part of the 1920s would be correct for this item. The barograph movement is of a twin-capsule aneroid design with a C-spring tensioning. Unfortunately, it is not signed, as one would expect on an item such as this, which was obviously made to be of great decorative value.

A very rare barograph by C. P. Goerz of Berlin, number 3621, is illustrated in *Fig.* 4.32. The wooden case, measuring 8¾ inches by 4½ inches by 13¼ inches high (22 x 11 x 34 cm), utilises a continuous strip chart, which does not need replacing, although the present chart was hand-made by the previous owner as a replacement for the original because of wear and tear. The design was patented, number 18320, in 1905 by Adolf Abraham near Hamburg. His patent is almost identical in design to this and also covers circular dials incorporating the same design.

The case leads one to think that this barograph probably dates from the 1920s but it could be earlier. The barograph capsules are behind the open dial, which has a needle to indicate air pressure. Working from this are two wheels, which carry a fine chain that has a loop on it at the top. Through this loop is fed a chain which can be seen laid over the chart. As the clock rotates the paper chart, magnets at the back hold the chain, which is repositioned as the pressure changes, so one always has the last seven days of

Fig. 4.31 Reverse of mechanism in *Fig.* 4.30 showing clock mounting and levers for dial barometer.

pressure recorded, with no need to change the chart. However, the clock, which appears to be of equally good quality, would require winding once a week.

Fig. 4.33 shows the heavy metal frame, which holds the mechanisms, removed from its wooden casing. The door does not have a catch or lock on it as usual but instead is unscrewed from the back and then opens as normal. Obviously, on display, there would be no need for any routine maintenance of ink or charts and the clock is wound through a hole in the case on the right-hand side. An earlier patent 10791, dated 1894, incorporated a similar idea by using a continuous chart with perforations on its margins driven by a sprocket-type drive from the top. This was by James Naylor Jr of Massachusetts, USA, but the charts have to be changed as no magnet is used and a line is drawn for pressure changes as in a standard barograph.

Fig. 4.34 is a French mahogany-cased barograph with 'PHNB' inscribed on the base plate. This stands for Pertuis Hulot Naudet Berthod which I think are the names of the founders of the instrument company that made it. The barograph is numbered 103 and the top of the case labelled 'Artaria Optn Fabt, Geneve – Maison Fondée en 1806', which is the shop that would have sold this item. Interestingly, a French firm by the name of Naudet still survives and I expect is a descendant of the Naudet involved in this company. The barograph has one glass panel to the front with a hinged lid, four pad feet, and has machine-cut joints. The case is identical in design to those

Fig. 4.32 A wooden-cased vertical barograph by C. P. Goerz of Berlin from the early 1920s.

sold by Richard Frères even down to the two plated hooks to the left-hand side of the case. The movement is of the usual aneroid design with a spring pulling apart the seven capsules. The tension arm has a counter-balance which slides for the optimum adjustment; the brass bar fixed to the top of the two tall pillars is inscribed 'Baromètre Holosterique'. This is quite an unusual movement, possibly dating from the 1890s.

Fig. 4.35 illustrates a barograph, measuring 13¾ inches by 8¾ inches by 7½ inches high (35 x 22 x 19 cm), in an oak case with flat glass; the moulded plinth has two strips of wood back to front forming the feet, the style of which was very common on later models and on nearly all barographs after

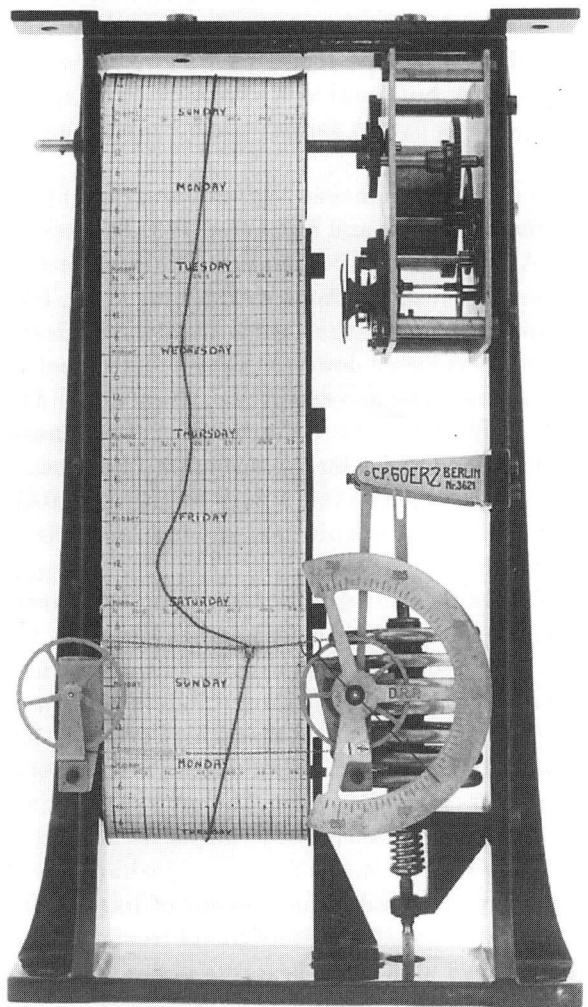

Fig. 4.33 Movement of barograph in *Fig.* 4.32 removed from its case, showing cast-iron frame, clock and bellows.

the Second World War. This particular barograph is inscribed 'A&NCS Ltd, Westminster', famous retailers of all types of instruments and equipment. The unusual mechanism of the barograph, which is supported on only two pillars instead of the usual four, can be seen in *Fig.* 4.36. This was perhaps an attempt to economise or vary the traditional design of the barograph. The two spindles necessary to transfer the movement from the bellows to the arm are arranged with brass brackets above the two pillars and are exactly the same arrangement as in a standard instrument. The adjustment for altitude is through a knurled knob between the two pillars; the arm is of the earlier sprung type and the four capsules are silver-plated. As only four

Fig. 4.34 A mahogany-cased French barograph, c.1890s.

capsules are used, the magnification is increased by having the connecting rod on top of the bellows very close to the fulcrum point on the main beam, taking the movement back to the arm spindle; return of the pen to the same position (when pressure is exerted on the capsules) is therefore not always guaranteed in my experience.

This design seems not to be particularly beneficial, and by magnifying the movement so excessively, the action of friction on the moving parts is also magnified and deadens the strength of the bellows. The clock is also unusual. *Fig.* 4.36 shows the gearing that is close to the base and there is a short spindle to locate through the clock housing. This is still the pre-1902 design and, with the lid removed (see *Fig.* 4.37), the mechanism can be clearly seen but much lower down in the drum than would be the case in post-1902 designs. The key, therefore, is made longer to enable easy winding when the lid is removed. Overall, I would date this item to around 1890. It has no manufacturer's marks so may have been made by one of the many smaller workshops turning out instruments at this time.

Apart from the obvious horizontally held bellows and the linkage changes, the barograph in *Fig.* 4.38 is actually of much the same design as all the others. It is by C. Plath of Hamburg and the inscription is dated 1922. There is a silver engraved thermometer between the clock and the bellows; the arm is of the tension type, and the case is in oak with bevel-edged glass. The adjusting screw for altitude is shown in the close-up

Fig. 4.35 An oak-cased barograph inscribed 'A&NCS Ltd', c.1890.

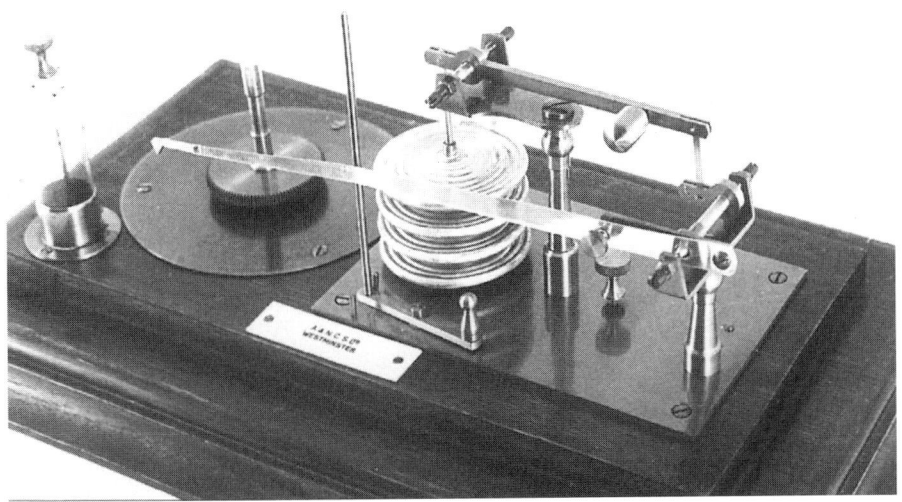

Fig. 4.36 Close-up view of the mechanism of the barograph in *Fig.* 4.35.

Fig. 4.37 Top view of clock drum with lid removed of the barograph in *Fig.* 4.35.

Fig. 4.38 Barograph with horizontal bellows in oak case by C. Plath of Hamburg, 1922.

Fig. 4.39 Close-up view of the mechanism of the barograph in *Fig.* 4.38.

view (*Fig.* 4.39) to the right-hand side between the pillars. This is indeed an interesting variation. I have heard of others of this design and *Fig.* 4.40 shows a small version in a typically French-style case with a half-size drum. The case measures 7 inches by 4½ inches by 5½ inches high (18 x 11 x 14 cm). These two barographs are plainly by the same maker as can be seen when comparing the movement in *Fig.* 4.41. The design of the brass pillars and the two decorative brass caps on top of the two taller pillars must be by the same firm. Whether this is indeed the German firm C. Plath or not I am unable to prove, but there is a good possibility, especially as inside the clock is the same trademark as in the German barograph featured in *Fig.* 3.44 by Tobias Bauerle and Sons (see chapter 3).

The barograph illustrated in *Fig.* 4.42 was sadly much damaged and required a great deal of restoration work. It utilises two clocks, one to turn the chart and one to operate a motor to adjust the position of the siphon tube. With the use of a governor, the siphon tube is allowed to move up and down, at the same time moving the pen along the chart. This is similar to the instructions of patent number 12476 dated 1902 by Edward Weston, which utilises electricity, but this particular barograph is earlier than 1902, and follows the design of Antoine Rédier, who died in 1892, so it can be dated c.1890 or, more likely, c.1880 (see Maxant 2000, p. 78). The bottom section, seen in *Fig.* 4.43, would have been enclosed in a glass case.

The barograph illustrated in *Fig.* 4.44 is very similar to the one shown in *Fig.* 4.42. It was made by Antoine Rédier but sold by Lund & Blockley, 42

Fig. 4.40 A small mahogany barograph with horizontal bellows in French-style case, c.1900s.

Fig. 4.41 Lid removed from the barograph in *Fig.* 4.40 showing mechanism.

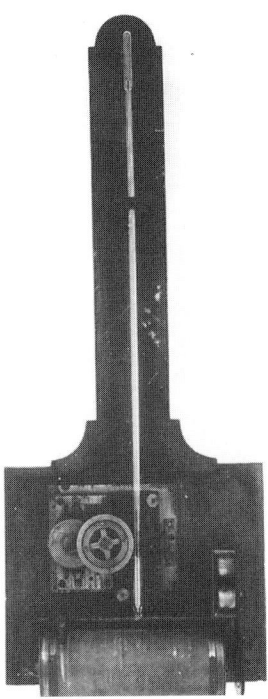

Fig. 4.42 Moving siphon tube barograph, c.1880.

Fig. 4.43 Close-up view of the mechanism of the barograph in *Fig* .4.42.

Pall Mall, London. It is a large piece, measuring 20¼ inches by 8⅜ inches by 42 inches high (51 x 21 x 106 cm), and housed in a mahogany case with fully glazed door and hinged access door to one side. Very few of this type of barograph survive. The large and complex movement clearly lost out to the more popular aneroid design we are used to seeing today.

Fig. 4.44 Mercury tube barograph by Antoine Rédier, c.1875.

5 General Maintenance and Repair

In general terms, barographs are quite easy to maintain, although one has to be relatively careful when handling them. Most barographs have a throw-off arm to move the pen away from the chart when changing the chart or winding the clock or when the barograph is in transit. Generally, for normal transit, it is a good idea to soak out any remaining ink within the nib so that it does not splash if vibrated.

The two most important features of a barograph are the clock and the mechanism. The clock should ideally be cleaned every five years, certainly every ten years, although we have had many reports of barographs working for thirty years or more. The clock, if used regularly, is a constantly moving piece of machinery and some care and consideration should be afforded it. Sometimes springs will break or become stiff. The bearings of the gears will need suitable oil administered by a qualified clock restorer. Do not, on any account, squirt WD40 or normal oil into these precious items as we have seen on a number of occasions. It might get a clock ticking for a while, but when the oil thickens the clock will be well and truly gummed up.

As with any clock, small amounts of dust will eventually enter the mechanism through minor openings, even though the glass cover is on it and the lid is on the barograph drum. The constant action of the gears will create some wear, and the oil, even the best, will get thicker. It is this oil that once thickened with metal particles becomes a paste and wears away the metal even more. This is why clocks should be carefully cleaned reasonably frequently. Fortunately, a competent clock restorer can usually repair even the most damaged clock. Unlike modern quartz clocks, which will generally have to be replaced, a mechanical clockwork clock, as used for over a hundred years, can usually be maintained.

Handle clocks with care: they have brass gears so the clock drum should not be dropped or roughly positioned onto the gears. When replacing, it should be slid carefully over the centre spindle and allowed to engage into the teeth gently. It is usual to position the clock with the chart on it approximately in the right time, and bring the clock up to the time indicated by the pen. It is not advisable to rotate this around continually as it wears the slipping clutch device and on a few occasions the clock will be found to be ticking but not revolving as well as it might. Regulation of the time is often a very difficult operation. Usually there is a small regulator lever visible alongside the winding key. Great care must be taken in adjusting the time because damage to the fine balance wheel can be expensive. In our own workshops, we try to get barographs to work within an hour a week, but we sometimes consider it acceptable for an antique barograph to be within two hours of a week. The backlash on the gears often gives you a slight difference, and the strength of the springs can sometimes mean that it runs at a different speed in the middle of the week from the end of the week, so the user should not generally expect a high precision, to-the-minute time-keeping piece. We have noticed that even quartz clocks suffer from a lack of accuracy due to the high gearing down of the drum.

If you consider that a revolution of the drum takes seven days, one hour is quite a small proportion of the chart. Also, the charts can expand and contract with moisture. There are many charts printed with different times and the original papers are often not available for old clocks. Although we produce several charts based on original divisions, it is seldom that the operator requires highly accurate timing. The barograph is more often used as an indication of approaching storms and it is most interesting to watch the drop of the arm on the barograph chart and the subsequent associated rise in advance of the weather improving. For much of the time, especially during summer weather, the barograph records an almost straight line around the chart and can sometimes be monotonous for several weeks. They are, nevertheless, important instruments to forecast the weather and are used by many people today, not only as an interesting decorative antique but also as a functioning instrument.

Most barograph clocks have a dust cover. If they have a glazed top, this should be kept in position to reduce the dust getting into the mechanism. Some clocks that have solid wooden topped cases do not have a cover.

The barograph pressure mechanism should be run dry and not oiled. The pivots should be free, and if the reader is fortunate enough to possess one by Negretti & Zambra that has jewelled pivots, these should not be over-screwed as damage to the jewels is quite likely. The mechanism should be fairly slack in terms of the pivots but not too loose.

If, over the years, ink has been spilled onto the barograph there will often be a number of stains, and we recommend that great care is taken when using wet ink so that damage is reduced. If the barograph has had severe misuse over the years or is extremely old, it is probably beneficial to clean and re-lacquer the brass with a traditional lacquer. The brass work should not be cleaned regularly. It is normal for these instruments to be lacquered or gold-plated, as with some of the Short & Mason items. These often last exceedingly well.

If the barograph has to travel by air, then it is recommended that the pin connecting the bellows stack to the lever be removed to allow the bellows to go up and down freely without stressing the arms and linkages. It is usually a simple procedure to replace this on arrival. Of course, if the destination is at a high altitude, adjustment to the arm will be necessary, probably by loosening it on its spindle and positioning it to its optimum to allow adjustment for normal altitude changes through the setting knob, which on many is positioned above the two main pillars close to the bellows stack. On some models, it is positioned on the base plate. On Richard Frères' barographs, it is usually operated by the other end of the winding key underneath the barograph.

Barographs should not be positioned where the sun can shine through a window or in any place close to a source of heat such as a radiator. Heat can damage the cases as well as discolour the brass more quickly and you will often find a recording difference on the chart due to heat, although some are partially compensated. We have seen a number of barographs in poor condition after being placed in a window for many years. The sunlight bleaches the back of the case, and the fitments and metalwork usually suffer. It is often nice to see barographs on a shelf at a higher level, and some barographs were originally sold with shelves, although generally these do not survive or follow the barograph.

The pen arms are often made of aluminium, although later ones used nickel silver and modern ones such as those made by Gluck have stainless steel arms. They should be carefully adjusted. The screw on the tensioning type arms should be adjusted so that the pen is only just touching the chart; indeed, a fraction away from the chart – such as a quarter of a millimetre – is usually acceptable as the ink will bridge the gap through capillary action. The drums are seldom perfectly round so there is usually a slight amount of difference as the drum turns around in a week. The lighter the tension arm is set, the less friction will be placed on the barograph chart and so the freer the movement will be. If your barograph regularly shows a series of steps, it may be that there is too much friction on the moving parts and the arm cannot move freely to record the pressure changes. On some barographs, occasional steps, caused by a combination of friction and the rate of pressure change, are inevitable, but they should not be the norm. If the arms are bent or corroded – the ends of aluminium arms often corrode – these should be replaced. The ink often reacts with the aluminium and after a number of years they corrode away. Aluminium was probably originally used as it is a very lightweight metal, but stainless steel has been used in more modern barographs for a number of years.

Pen nibs should be kept clean and they can be replaced when necessary. I expect a new pen, which today is usually made of stainless steel, to last many years and more often they become damaged rather than corroded. Traditionally, a bottle of ink was placed inside the barograph case with a little brass dipper, though these have often corroded away at the ends. I generally think it is a good idea to have the ink bottles cleaned and in position but not to use them. Today's modern plastic dropper bottles are so much easier to dispense ink that we suggest that these are used instead and the barograph bottle kept mainly for show.

If you have a barograph with a drawer, these often have holes in the base. These are to ease removal of the chart by pushing your finger underneath the drawer to lift the charts up. The chart drawers are usually designed with two compartments. Last week's chart can be placed in the front on the top of the pile of used charts so that you can, if the chart has just been changed or within a day or two, open the drawer and see last week's pressure trend. New charts are therefore kept in the back of the drawer in the second compartment. Whilst it is an economy to allow the chart to be used for several weeks, after a while it becomes untidy and when you experience the sudden drop of an interesting pressure change (such as occurred in Britain in 1987 when the trace nearly went off the bottom of the chart) you may not have a clean chart in place to record this, which can be disappointing.

To operate the standard type of barograph, one should wind it once a week, change the chart and check the ink. I often find that a drop or two of ink will last several weeks so it is not always necessary to put in fresh ink. A bottle will often last several years. The ink used is generally a purple/blue colour; thermograph ink is usually red and hydrograph ink green. Modern fibre-tip pens can be adapted for use on old barographs, if required, but I think this is a little insensitive to the instrument and I do not like to see plastic on the end of an old barograph arm. It is in fact usually more expensive to use disposable pens as they cost several pounds and may only last 12 months (although many we sell have been working for three or four years). If the nib becomes choked after several years' use, the dried ink can be scraped out carefully, or if the nib and arm can be removed, the nib can be washed out with methylated spirits or warm water (see chapter 6 for further advice on nibs and ink).

For altitude, barographs should be adjusted in the same way as dial barometers so that they are reading a figure that corresponds to a sea-level reading. Therefore, when you hear the reading on the radio, or if you phone an airfield or the Met. Office or compare with someone else's accurate barometer (not a stick barometer which reads actual pressure), you can adjust with the adjusting screw or knob so that the barograph reads correctly as if it were at sea level. It is important to adjust barographs correctly, otherwise, during highs or low pressures, the arm may exceed the height of the chart and not give a true recording. Nearer the poles, the greater the pressure change, and so a barograph is of even more benefit the further north one is. Barographs with a hinged gate arm should be adjusted so that it swings lightly to the chart and is not heavily inclined; this way friction is reduced. Indeed, the hinged arm is far superior in operational terms than the earlier tension type of arm.

Providing general common sense is used, care is taken of the barograph and the clock occasionally cleaned, it should last a lifetime – and beyond. We have had many people bring their father's and grandfather's barographs in for restoration – a true testament to the engineering skills of the craftsmen who built these wonderful instruments. The same cannot be said of electronic instruments: ten years can be a long time for them to survive. Despite some owners' neglect, barographs can almost always be restored to their former glory.

Possibly the most delicate and difficult item to replace is the bellows. Old barographs have a varied number of capsules screwed together to make the bellows stack and modern barograph bellows are often too sensitive to use as replacements. The would-be restorer would be well advised not to try to mend barograph stacks or tinker with them. It is sometimes possible, however, to remove one capsule, if it has blown (the vacuum has leaked), and replace the longer linkage connection and maintain an operational barograph. Removing two capsules will often mean insufficient movement to record true pressure changes (see chapter 6 for more on bellows).

Ink for barographs is of a special type (see chapter 6) and ordinary ink should not be used as it dries too quickly. If any ink is spilt on the woodwork or metalwork when refilling, this should be wiped up immediately with a tissue. It is mildly corrosive, although it is very easy to clean up a small drop or two if done quickly. It is not corrosive like acid but an accumulation of ink will begin to corrode the brass and pit it and, of course, stain the wood, which is unsightly at the very least.

Many barographs that have charts held on with clips should be monitored to make sure that they are allowing the pen to run over them. If the clips have been badly bent through misuse, the nib will sometimes catch on them at certain levels of pressure and pull the nib off or create some damage to the arm. I must admit to preferring the adhesive type of charts, which were patented by Short & Mason in 1902, as it makes the traverse of the pen so much easier. These should be laid the correct way so that the pen falls off the edge of the paper rather than being caught in the advancing edge if the paper is positioned on the drum the wrong way. Most charts are printed with 'place under other end' on the appropriate side for this reason. Charts should be changed on Sunday or Monday (according to which chart is used) and Short & Mason recommend that this should be between the hours of 7 and 12, being the period duplicated on the charts for this purpose.

Basic Repair Tips

One of the most common problems we encounter in the workshop, other than clock issues, is that the linkages that magnify the rather small up and down movement of the bellows to the arm have been damaged or altered incorrectly – even sometimes oiled. The barograph mechanism should not have any oil on it, and all the linkages that have tapered pins through to connect the rods to levers and so on should be free and not twisted. If a pin is rusty, it can create tension and cause problems and incorrect readings. Each pin should be smooth enough to allow the link in the middle to move freely. Tapered clock pins are not essential; indeed, some barographs I am sure used thin wire. The linkages are held under slight tension by the weight on the arm from the bellows stack so even a loose link will work well. The tapered pin should fit tight in the brass bar that is split and goes around the link but the link must be free moving.

Once all the linkages have been checked carefully – a magnifying glass can be helpful here – look carefully at the angle of the slot that the links fit into: often, for no apparent reason, the brass rod with the slit in has been moved so as not to be upright and the linkage is therefore straining to move at such an angle. Even a very small angle can restrict the movement of the arm. *Fig.* 5.1 shows a slotted brass connecting rod twisted and causing a lot of friction on the linkage. Just a small amount of extra friction can cause a sticky movement of the pen. If the hole in the link is too small or rough, then carefully running a suitably sized reamer in it should make sure there is plenty of free play on the pin when assembled.

The pivots that the axles are held between are normally pointed screws. These need to be adjusted so there is plenty of free play but not excessive; if too tight, nothing will work well. It is these that some inexperienced restorers oil, but do not be tempted. On the Negretti & Zambra 'jewelled' barographs the end of the screw has a small concave jewel in which to rest the pointed rocking. If overtightened, the jewel can crack and cause problems. I have replaced some jewelled screws over the years and they are not easy to obtain.

Another frequent problem is faulty bellows (see also chapter 6). There are a few replacements available, but replacing them is not straightforward – if you want the barograph to respond correctly to pressure changes, of course. Similar to changing the engine in a car, you should have the right type, but with barographs the original manufacturer has normally long ceased trading and exact replacements are no longer available. In my experience, the modern replacements move much to much, so when we have to change a bellows on an old barograph we have frequently needed to unscrew three or four capsules, drill a hole through the female thread and tap into the other side of the capsule. In the back of the male thread, place a small 10BA countersunk screw to permanently fix the capsule and then screw it back on the stack. This makes that capsule decorative only, but reduces the amount of up and down movement so as to make the pen rise and fall more accurately on the chart.

To ensure there is correct movement, it is advisable to place the barograph mechanism in a pressure chamber to see how accurate it is. If you do

Fig. 5.1 Close-up view of twisted flat linkage causing friction.

not have access to a vacuum chamber, you could wait and observe the pressure changes naturally and then make adjustments once enough change has been compared with another bellows, but this is very time consuming. If the bellows you fit are close enough to correct accuracy, or the old bellows have failed a little with age and are under-performing by a small enough amount, it can often be possible to adjust the levers so as to make more (and sometimes less) magnification. The far end of the bellows-connecting rod may have two holes that can be used to make changes to the magnification of the pressure. The further away from the bellows that the link is positioned, the more movement of the rod is transferred to the arm-connecting rod below. The arm-connecting rod may similarly have additional holes (see *Figs* 6.16 and 6.18 in chapter 6).

The reverse of the bellows rod, the arm rod magnifies the bellows movement more if the link is moved closer to the rocker rod that the arm is attached to. On some models, which tend to be older and in my opinion of better quality, there is a small screw adjustment on the arm rocker rod (see *Fig.* 3.48 in chapter 3) which does the same job as using different holes on the arm rod and, more importantly, allows the adjustment to be much finer.

On barographs where more magnification is needed but there are no additional holes, it can often be possible to elongate the slit in the arm rod and drill a few more holes or drill more holes near the pen end of the bellows-connecting rod. On the rare occasions that a barograph is reacting to pressure changes more than it should, you would naturally drill holes or move the link the opposite way. The only other way to adjust the amount of movement of the arm to respond to pressure changes is to slide the bellows nearer to the clock to lessen the movement or away from the clock to increase the movement. *Fig.* 6.19 in chapter 6 shows the underside of a barograph where the bellows stack is screwed to the base plate in a slot. This enables calibration when it is being made or when needed if the bellows are not working as well as they should.

Apart from incorrect assembly of the barograph, the main reason bellows sometimes become under-responsive to pressure changes is that the vacuum can very slowly degrade over time – not a leak in the usually accepted way, but molecules of air seeping through the metal (yes, I believe that can happen in time and have spoken to scientists on the subject on occasion). This leakage is small but can affect the operation of the bellows, and if small enough some adjustment can be made to increase sensitivity by one of the means aforementioned.

Apart from correct setting up of the linkages and rods, and adjustment for magnification of pressure, most other repair work is down to common sense. Case work normally requires the same cabinet work as other antique furniture, and polishing of the case is best done with French polishing, a specialist subject in itself. Broken bevelled glasses will need making specially by grinding – not a job for the general restorer as it has to be done well and with special grinding equipment. Clockwork is for the clock restorer. However, with the above information, a practical mind and skilled hands, you should be able to undertake many tasks needed to maintain and repair some of these fine instruments. But, as always, if in doubt, use the services of a skilled professional as a mistake can be costly or impossible to rectify.

Metal Cleaning and Finishing

Most barographs are made of brass and some steel parts. With age, these naturally deteriorate. In extreme cases, the steel parts can become rusty, although seldom a deep rust, and the brass can become oxidised to a dirty colour that detracts from the look of the instrument, although this will almost never affect its operation.

With steel, there are often two types of finish: the brass pillars may have a burnished and lacquered steel bar across them, and the small bracket that protects the back of the arm on models like the early Richard barographs is also often made of burnished (polished) steel. These should simply be cleaned and polished until a bright steel finish is achieved, then a coat or two of clear lacquer should be applied to protect them.

Sometimes these pieces, as well as a variety of screws, are finished with heat blueing. On lightly aged blued screws, one can often wipe them over with a light oil to freshen them up and the thin oil coat will protect them for awhile. You could lacquer them if required, but the blueing treatment is itself a mild protection against rust – for a period. Barographs are generally kept indoors in dry conditions and have a glazed case over them so the steel does not often rust much. To re-blue screws, one needs to polish the head well: the smoother the polishing, the more even and blue the finish is if done correctly. If a lot of screws need re-blueing, then I use an electric kiln to control a temperature rise to approx 195 degrees centigrade, but each kiln can vary in its accuracy and a little testing is needed to ensure a suitable colour. A lower temperature will produce a lighter finish; a higher will produce a darker finish, but too high and the colour is very poor and needs doing again.

We use this process to blue hundreds of screws at a time once the heads have been cleaned. But if only an odd screw or two are needed, then very careful holding in a gas flame until the right colour is achieved and then plunging in oil to cool and seal the metal can work well, but be careful as the colour changes very quickly and you can soon over-cook them! A few trials before doing the screws needed would be wise.

With brass, it is sometimes the case that the whole barograph needs dismantling – every little screw and piece – then cleaning individually with brass cleaner and wire wool to remove the old lacquer before burnishing by hand with a cloth. Avoid polishing with a burnishing mop as the parts are so small and nearly impossible to hold and you can so easily over polish and remove metal on the corners. Once cleaned and then wiped over with thinners to remove any traces of polish, they are then ready to lacquer.

With so many pieces, it will be best to make some system to hold them. We use a variety of pointed wooden sticks (see *Fig*. 5.2) in a wooden block and suspend those parts that have suitable holes in them; for screws that only need the heads lacquering, we use a cardboard box with holes punched in it to hold the screws (*Fig*. 5.3). Sometimes one has to be inventive and wrap wire around parts (see *Fig*. 5.4).

It is possible to lacquer with a brush but you will need to be very careful when applying the lacquer and use a good-quality brush. Since I invested in a small pencil spray, it has become somewhat easier to get a more even coat without dribbles of lacquer. A number of brass lacquers are available from material suppliers. We use a clear satin finish as gloss really does not suit these instruments in my opinion. I would not recommend using just beeswax as it wears off and seldom lasts as long. Over the years, we have often used a gold-coloured lacquer that we make ourselves, but there are a variety of coloured lacquers offered by clock parts specialists which should also suit. If using a coloured lacquer, then spraying is far easier than attempting

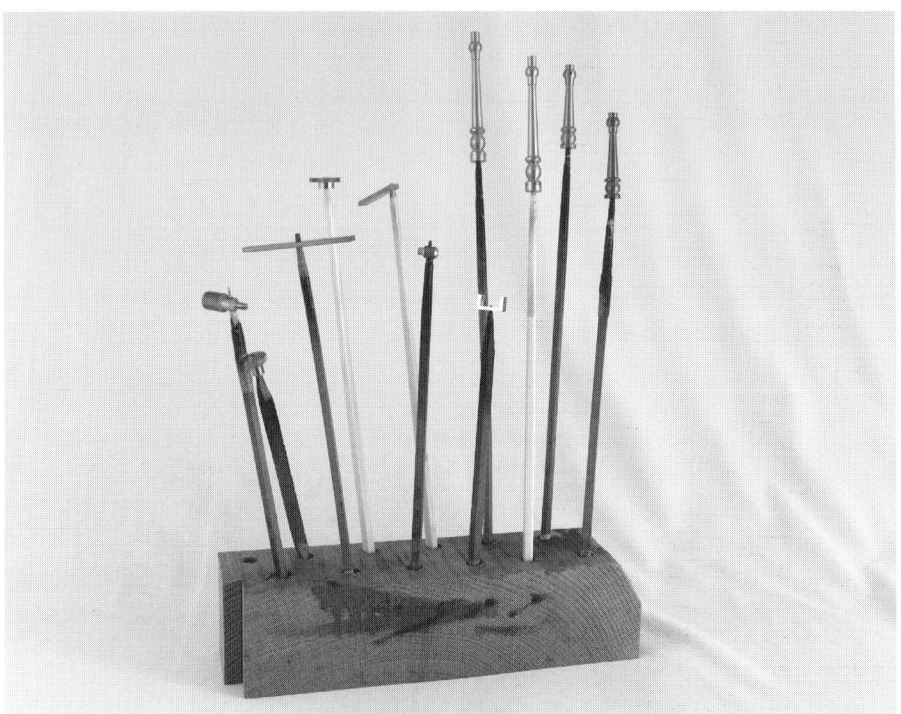

Fig. 5.2 Holding brass parts on sticks to lacquer.

Fig. 5.3 Screw heads held ready to lacquer in cardboard box.

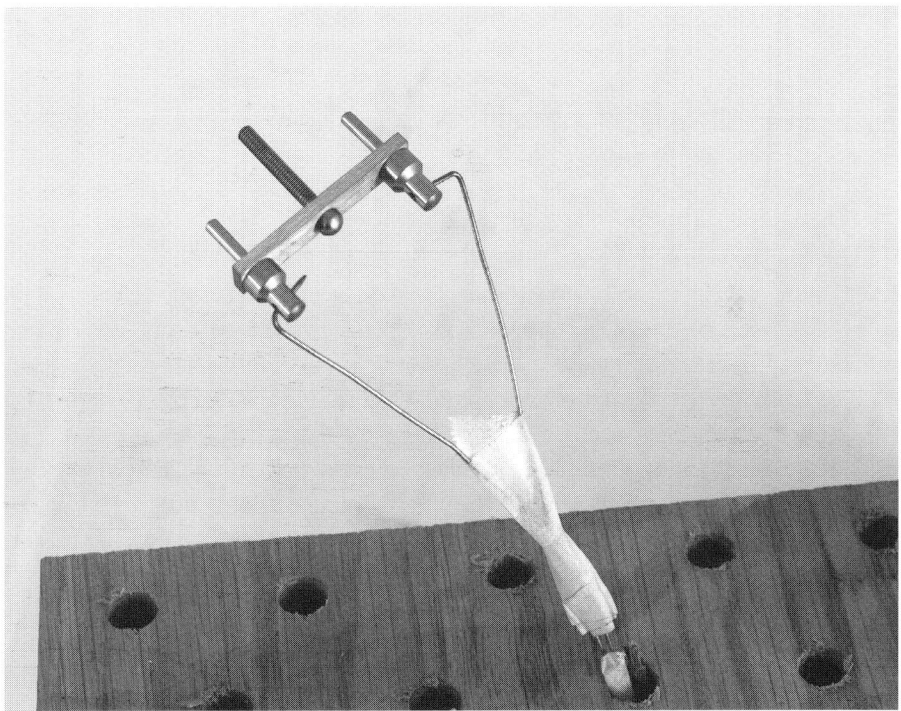

Fig. 5.4 Wire used to hold parts for lacquering.

to brush on the lacquer, which is possible but requires high skill.

The bellows should not be cleaned as the other parts. Care is needed if it is one of the types with soldered evacuation nibs: if you bend or break one of those you risk letting air in and ruining the vacuum. Often they are silver plated, and with age this has discoloured, though some may be tinned. Either way, avoid aggressive cleaning. Generally, a careful wipe over the top and the removal of dust between the capsules should be enough. Perhaps re-lacquer the top to prevent it from becoming more discoloured in the years to come. The linkages joining the connecting rods together should not be lacquered, and any lacquer that may have got between the slots of the connecting rods should be cleaned out. Indeed, it is often sensible to fold over some fine wet and dry paper and run this through the slot to ensure it is clean and smooth so as not to cause any more friction than normally occurs.

Be careful if the barograph has a blued finish to the brass plate and sometimes the clock drum, as some were made like this. I have not yet found a way of re-blueing – mainly as it has not been necessary. A careful light wipe over and re-lacquer is often all that is needed. Once you start cleaning such a brass plate and reveal the brass, you will end up having to sand the whole plate back down to brass to obtain an even brass colour.

If the pen arm is damaged, you may have to make a new one. We use thin aluminium and drill the holes as needed. Other types were used: some can be straightened if not too badly bent. Fitting a new nib is sensible when doing such extensive work as well.

When all the parts are cleaned and lacquered as required, you should be ready to re-assemble the instrument. For the novice, it is probably now that you wish you had taken pictures of the instrument from all angles to see how to re-assemble it all! One needs to be very experienced to be able to re-assemble all the parts without thinking about where each bit goes. A picture will help to start with, and when a little more experienced, a sketch or two of some of the assemblies is all that may be needed. As a general rule, I replace the pins securing the linkages to the connecting rods.

Before screwing the mechanism back on the wooden case, it is best to check that the item is working well again in a pressure chamber. Without a pressure chamber, testing how well a barograph is working is much more difficult. One needs to check against another known barometer over a period of time when the pressure changes are sufficient to be able to compare – and this can take many months, of course. One simple way to ensure that all the linkages are working well is carefully and lightly to touch the top of the bellows and depress it – a gentle tap is needed and this should set the pen arm bouncing up and down a few times. If it appears sluggish and stiff, then there is something wrong (unless it is a dampened variety for use on board ship, of course). If the bellows are good and you have followed all the above instructions and have assembled the mechanism correctly, it should work well, but with this type of work it is so easy to overlook one small aspect. You will therefore need to check the assembly very carefully – I find using a magnifying glass helps considerably.

6 Bellows, Nibs, Charts and Ink

Bellows

By studying some of the surviving catalogues of instrument makers who sold barographs, we get a feel for the development of this instrument. After Negretti & Zambra's early catalogue selling 'self-recording barometers' (see chapter 1), we come to their next major catalogue which was compiled by Major Frank Short about 1928–30, according to Jack Noble, a former employee. A hand-written note by Jack Noble in the front of my copy records that Major Short took over management of the Half Moon Works and signed documents as works manager. The catalogue that he compiled, *Engineering Instruments and Industrial* (List E5), still calls what we know as barographs 'recording barometers', although precision recording barometers are also mentioned as 'microbarographs'.

We know from patent 198838 of 1922 that bellows began to be made without soldered side edges, and Negretti & Zambra's List E5 catalogue describes some of these new bellows (or 'diaphragms' as they called them). *Fig.* 6.1 illustrates the 'old type of diaphragms' (left) and the 'Negretti & Zambra type of diaphragms' (right) which had a number of merits over the earlier types, as they state on p. 31 of their catalogue:

> The chief merit of the Negretti & Zambra diaphragms is that internal friction caused by the leaf springs is eliminated. The amount of this friction in the older instrument is frequently considerable and is apparent by 'steps' in the trace with a slowly rising or falling pressure. This defect is also obvious when an attempt is made to adjust the instrument with precision to standard pressure.
>
> The Negretti & Zambra diaphragms are formed of a material with high elastic property, and the movement of the diaphragm is on either side of its horizontal plane. The diaphragms under these conditions give practically even increments of pressure, and are so proportioned to operate well within their yield point. In the older type of diaphragms, the diaphragms were frequently formed of an imperfect elastic material, and with the fitting of internal springs, errors, due to hysteresis, friction, creep, etc., were often very marked.
>
> Compensation for temperature is made by leaving a cer-

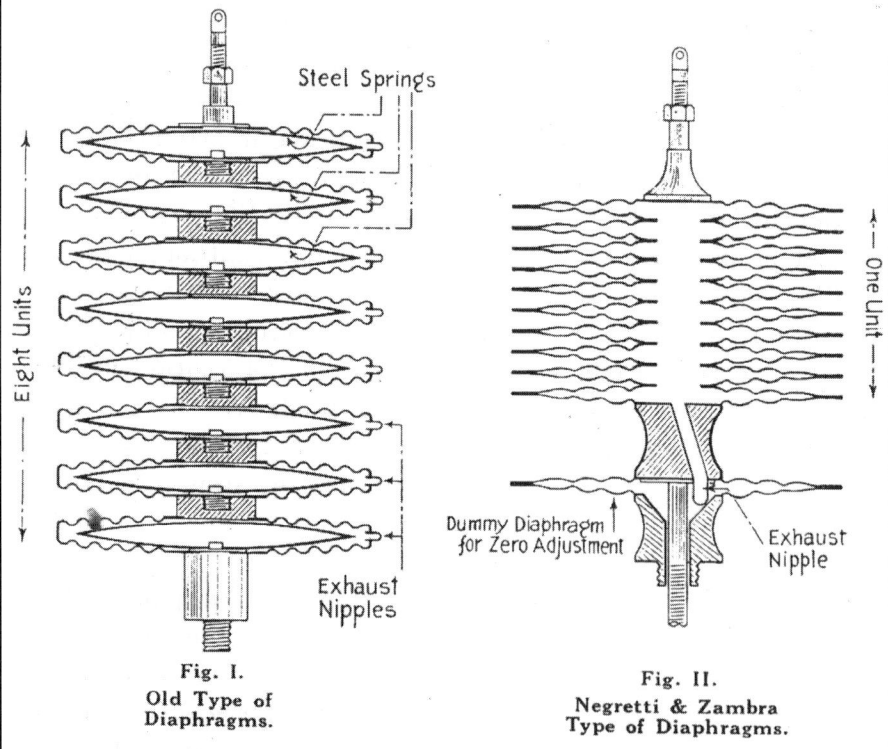

Fig. 6.1 Negretti & Zambra's new type of bellows ('diaphragms') from their 1928–30 catalogue (List E5).

tain amount of air in the diaphragms. In the older type of diaphragms this was done to each diaphragm separately, but in the Negretti & Zambra movement the whole unit is compensated together, thus ensuring greater accuracy.

In the older types of recording barometers the diaphragms were mounted on a leaf type spring as those of Richard Frères. An adjustment screw was provided which deflected the spring with the diaphragms mounted upon it. Apart from the coarseness of adjustment, the main defect was that the diaphragms were not adjusted up and down in a true vertical axis. In the Negretti & Zambra recording barometers this defect is overcome, as the diaphragms are mounted on a dummy diaphragm, which acts as a semi-rigid support, flexible in the vertical direction only.

The spring on the underside of the bellows of Richard Frères barographs, mentioned in Negretti & Zambra's catalogue description above, can be seen in *Fig.* 6.2. The screw has a square pin on the end which, when on the case, is accessible through a small hole in the base of the wooden case. The two screws on the left of this rather thick brass spring secure it to the base plate; the screw on the far right of this secures the bellows to it. By screwing the square pin with the key provided (or similar size if lost) the whole bellows can be moved up or down to adjust for altitude.

Fig. 6.3 illustrates a Negretti & Zambra barograph with these bellows or diaphragms. It has bevelled glass to five panels and the case is longer than the usual type of barograph case. The throw-off arm, also shown in *Fig.* 6.4, operates just behind the pillar. The typical Negretti & Zambra rubber eye-dropper ink dispenser top is shown in *Fig.* 6.5; it also has their later trademark of a stylised 'N&Z', c.1950.

Most Negretti & Zambra barographs of this design are numbered individually (this one is R/36140), and the base plate is engraved 'Regency Jewelled Movement'. The adjustment for altitude screw is also engraved ('Set Pen'). The jewels are in the four pillars where the axles move. Each pillar has a piece of threaded brass which screws into the pillar and is set almost level with the inside of the pillar. In the end of this brass screw is a small jewel (semi-precious stone) with an indent into which the pointed end of the axle rests, and, if set correctly, supposedly reduces the friction when the axle moves a little. I doubt if the gain is really significant, but it is a nice addition to the barograph. It is very important not to over-screw the jewelled studs as you can crack the jewel and I can assure you that these days they are not so easy to replace. Once the jewelled studs are in place, a finishing screw is screwed into the outside of the pillar. These can be seen in *Fig.* 6.6 which also shows how some calibration adjustment can be had by adjusting the length of the short rod which connects the long rod from

148 *Bellows, Nibs, Charts and Ink*

Fig. 6.2 Brass spring on underside of bellows of a Richard Frères barograph.

Fig. 6.3 An oak-cased Negretti & Zambra barograph with their 1930s' style bellows.

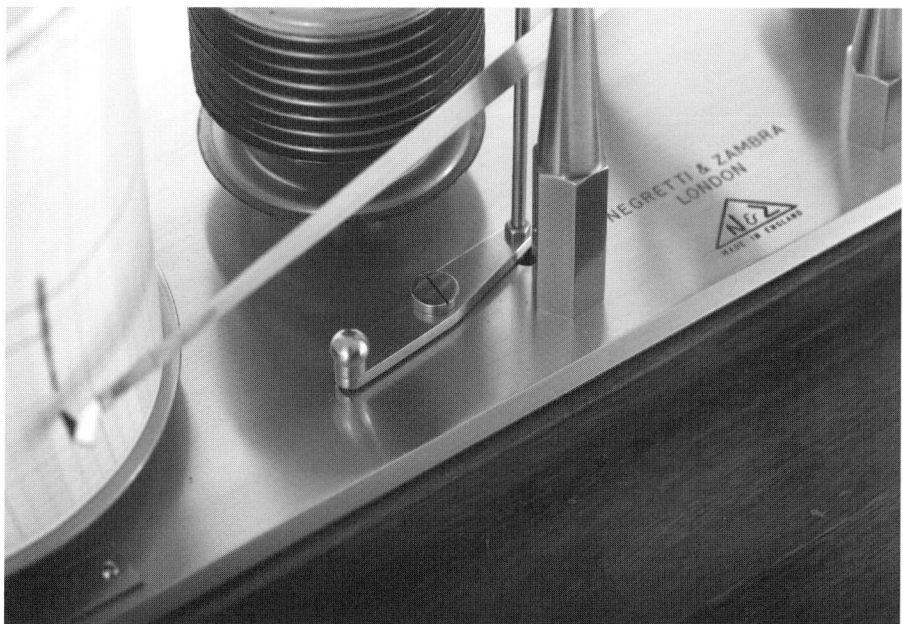

Fig. 6.4 Close-up view of barograph in *Fig.* 6.3 showing the mechanism.

Fig. 6.5 Original ink bottle with eye-dropper top in Negretti & Zambra barograph.

Fig. 6.6 Adjustable short rod on Negretti & Zambra barograph.

the bellows to the rocking axle held by the jewels.

It is without doubt that most (although not every one in my experience) of these 'Regency' barographs by Negretti & Zambra were very well-made instruments. This model used an early Gluck clock as shown in *Figs* 6.7 and 6.8. There is, interestingly, no counter-balance for the arm. A counter-balance weight, which is so common on old barographs, places a little pressure on the arm so that it is always slightly weighed downwards, thus ensuring that the nib falls back into position better. If, for example, the movement was well balanced without the weight, then you might expect to see some very minor fluctuations in the line when the pressure changes ever so slightly – just how much is not certain. On more modern barographs, this counter-balance tends to be positioned on the back of the short arm connecting rod (see *Fig.* 2.6 in chapter 2).

Variations can be found among counter-balance weights. *Fig.* 2.21 (in chapter 2) illustrates an interesting ball shape hanging from the main connecting rod. *Fig.* 6.9 shows a common style of weight screwed to the main connecting rod seen on many Richard Frères barographs. A similar one is on the dial barograph shown in *Fig.* 6.10. *Fig.* 3.18 (in chapter 3) shows the counter-balance weight on a Met. Office microbarograph dated 1938. On our own Merton barograph, we use a style of counter-balance that I liked from an old barograph we worked on which is a ball shape that can slide

Fig. 6.7 Details of common Gluck clock on Negretti & Zambra barograph.

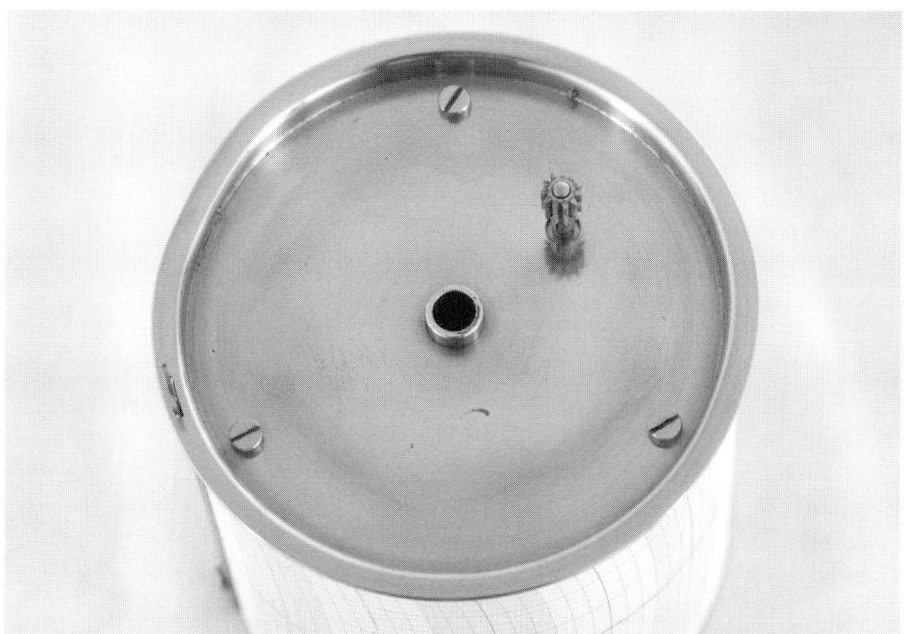

Fig. 6.8 Underneath of clock drum on Negretti & Zambra barograph.

Fig. 6.9 Counter-balance seen on many Richard Frères barographs.

Fig. 6.10 Counter-balance on a dial barograph.

along the bar to balance the weight more accurately (see *Fig.* 6.11).

Fig. 6.12 from the Negretti & Zambra catalogue also clearly illustrates the new type of bellows with its dummy adjusting capsules as previously described. These were available with daily or weekly clocks: with either, they cost £12 10s. Also clearly shown is the slightly unusual type of pillar used to support the mechanism: the two long and two short pillars are turned from a hexagonal bar, a notable identification of this type of Negretti & Zambra barograph of the period. Whilst Negretii & Zambra bought their early items from Richard Frères, and certainly bought many from Wilson Warden & Co. after the Second World War, I believe that in the 1920s and 1930s they were producing many of these instruments themselves, although almost certainly the clocks would have been bought in.

Negretti & Zambra's meteorological instruments catalogue List M4 dated 1950 illustrates very similar items. The same barograph shown in *Fig.* 6.12 was then on sale for £29 3s. A precision recording barometer or microbarograph, number 5090, is illustrated in *Fig.* 6.13. The movement is operated by four sets of bellows and cost £40. The metalwork was nickel-plate finished and the charts were 5.8 inches by 6.2 inches. *Fig.* 6.14 shows a slightly different dial recording barometer, which was on sale for £15, in an oak or mahogany case.

Fig. 6.15 shows the top linkage of the bellows which often has more than one hole into which the connecting rod can be located. This allows for variation in the height of the bellows and is very useful to aid adjustment when needed. The illustration also shows that the end of the connecting rod has two holes, allowing more movement and adjustment when calibrating in a pressure chamber. *Fig.* 6.16 shows the other end of the connecting rod that is linked to the arm axle with an extra hole. This also allows more opportunity for adjustment when setting up the barograph. Sometimes it is necessary to drill extra holes to make more movement when an old bellows has become slightly less responsive. The nearer to the arm the linkage, the more magnification is made. *Fig.* 6.17 shows a flat link that has a variety of holes. This can be useful to ensure that the barograph is set up so that the connection rods are roughly positioned in the mid-range – not too high or too low – to allow more range when adjustment for altitude is required. *Fig.* 6.18 shows the short connection onto the pen axle with additional pinhole. The variation of length here makes a great difference to the range of the bellows and I have often needed to drill extra holes to magnify the range due to bellows becoming less operational over many years.

Another method of altering the magnification or, to use the correct terminology, calibrating the barograph, is by altering the position of the bellows on the base plate. *Fig.* 6.19 shows the underside of the barograph with a slot in which to locate the bellows. By moving the bellows nearer or further away from the primary axle, and adjusting the connection rods

Fig. 6.11 Sliding counter-balance used on our Merton barograph.

Fig. 6.12 Negretti & Zambra's 1928–30 (E5) price list of oak-cased barographs incorporated their new diaphragms.

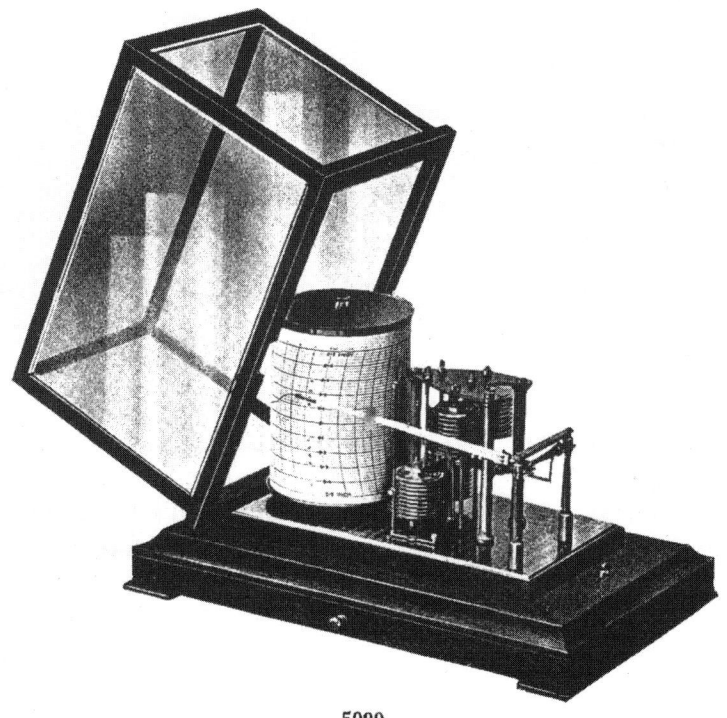

5089 **Precision Recording Barometer,** range 28" to 31", with 5" scale measurement. Chart 5.8" × 16.2". Movement, operated by 4 sets of diaphragms (see page 30-1), fully compensated through a link mechanism on spring struts and jewelled pivots to avoid friction and backlash. Large control on pen arm and open range chart. Nickel-plated finish in oak case with drawer and bevel glass panels. Complete with 100 charts and ink. Weekly or daily clock   **£34 10 0**

5090 **Precision Recording Barometer,** as above, but giving a record of − 0.5" to + 0.5" barometer, *i.e.*, 1" on the chart with 5" scale measurement. If the barometer reads 30" when the instrument is being set the pen arm is adjusted to 0 on the chart and the instrument records a rise or fall of 0.5", *i.e.*, from 29.5" to 30.5". The instrument may be set to any barometric reading within the range of 28" to 31". Complete with 100 charts and ink. Weekly or daily clock **£40 0 0**

Fig. 6.13 Negretti & Zambra's precision barograph from their 1950 catalogue (List M4).

Fig. 6.14 An open half-dial barograph illustrated in Negretti & Zambra's 1950 catalogue (List M4).

accordingly so that the bellows are upright, the barograph can be calibrated when being assembled with the use of a pressure chamber. This is the method I generally use when making our Merton barographs, but many old barographs do not have this facility.

Nibs

The basic principle of a barograph is to record the pressure of the atmosphere over a period of time, normally on graph paper. This recording has been done in a number of different ways. The French barograph by Breguet of 1867 (see *Fig.* 1.12) utilised smoked paper. Alexander Cumming's barographs (see chapter 1) used pencil to record on vellum charts. The Kew station barograph and a few other types used photographic paper to eliminate the friction of a recording pen, which is inevitable when a nib is in contact with paper even though the friction is often very minor. The first three patents listed in the Appendix indicate pressure recording using pen-

Fig. 6.15 Holes in the bellows linkage.

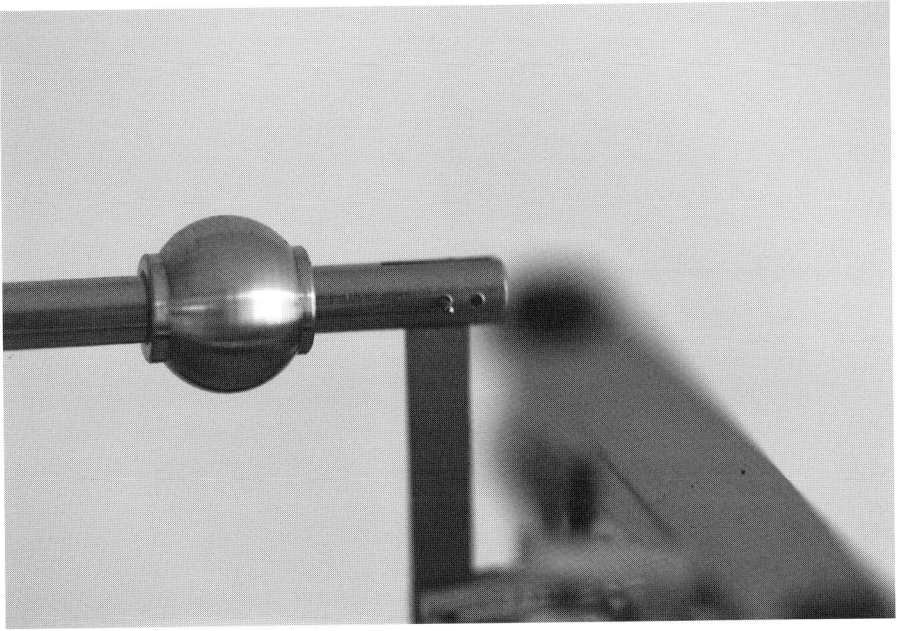

Fig. 6.16 Holes in end of the connecting rod.

Fig. 6.17 Holes in flat linkage.

Fig. 6.18 Holes in pen axle connecting rod.

Fig. 6.19 Slot in base plate to allow adjustment of bellows position.

cil on a paper drum. The fourth one, number 1680 of 1871, makes a trace by dots through carbonised paper. In *The History of the Barometer*, Middleton (1968, p. 288) quotes a description of Robert Hooke's 'weatherwiser', which had a hammer striking the punches once every quarter of an hour, so we can assume that holes were punched into a paper chart to record pressure changes.

The most noticeable step forward in recording the trace on barograph paper was, I believe, by Richard Frères, who patented a number of improvements, including a nib to hold ink, which is clearly seen in the patent of 1880 by Jules Richard. This design became the standard for many decades, and we still supply this style for old and some new barographs. *Fig.* 6.20 from Negretti & Zambra's 1928–30 instruments catalogue (list E5) gives some notes on recording pens, called the 'boat type'. These notes may prove useful to anyone trying to set their pen, which can from time to time prove troublesome. Adjustment of the points can give a wider or narrower recording line.

It is possible that further patents for pens may come to light. There are several different varieties but these are generally of quite late production. After the Second World War, a capillary-type pen was used, and in the National Meteorological Office Library and Archive (CIR-2028, INS-28, 30 October 1951) there is a circular from the Meteorological Division of the

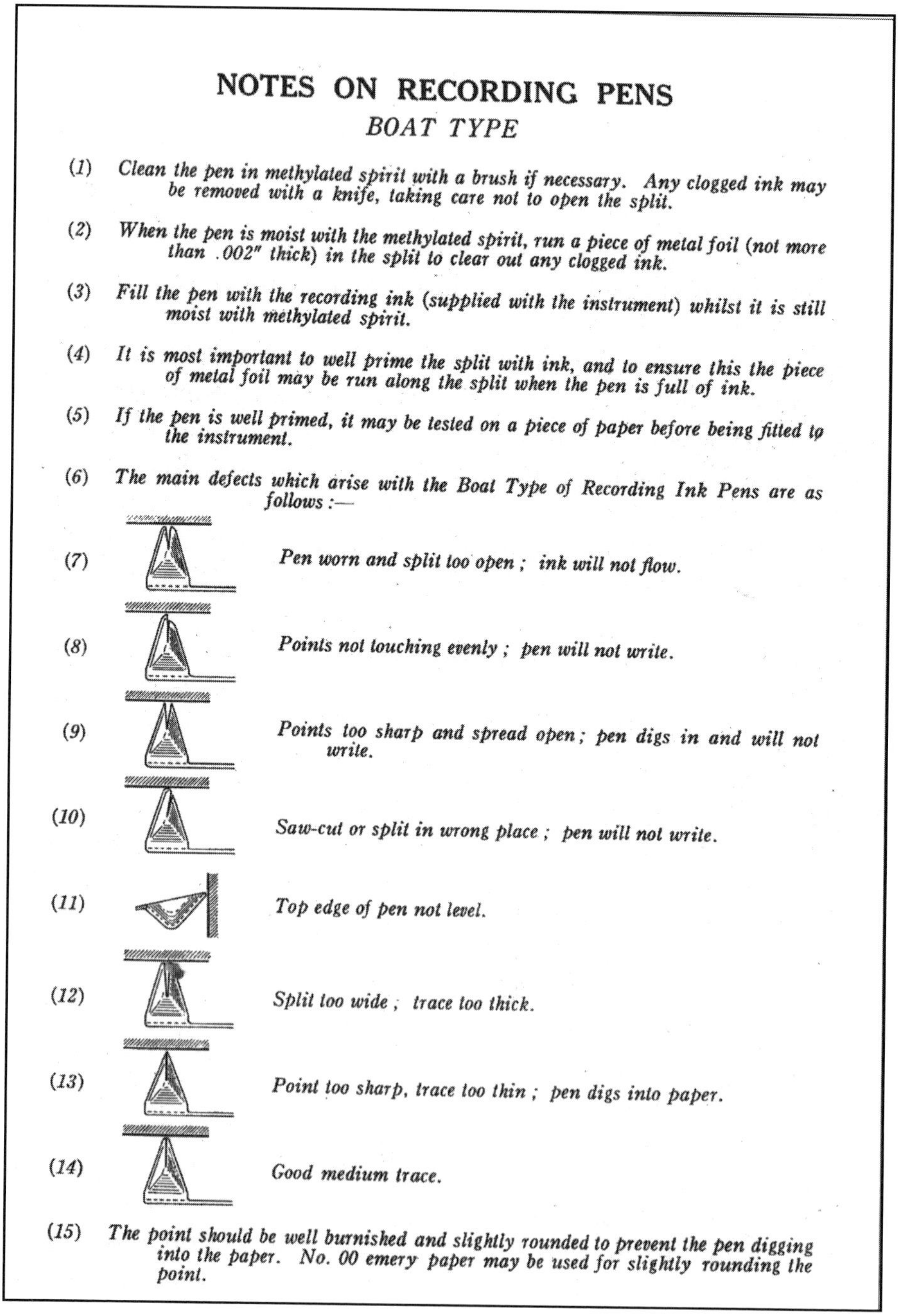

NOTES ON RECORDING PENS
BOAT TYPE

(1) Clean the pen in methylated spirit with a brush if necessary. Any clogged ink may be removed with a knife, taking care not to open the split.

(2) When the pen is moist with the methylated spirit, run a piece of metal foil (not more than .002" thick) in the split to clear out any clogged ink.

(3) Fill the pen with the recording ink (supplied with the instrument) whilst it is still moist with methylated spirit.

(4) It is most important to well prime the split with ink, and to ensure this the piece of metal foil may be run along the split when the pen is full of ink.

(5) If the pen is well primed, it may be tested on a piece of paper before being fitted to the instrument.

(6) The main defects which arise with the Boat Type of Recording Ink Pens are as follows :—

(7) Pen worn and split too open ; ink will not flow.

(8) Points not touching evenly ; pen will not write.

(9) Points too sharp and spread open; pen digs in and will not write.

(10) Saw-cut or split in wrong place ; pen will not write.

(11) Top edge of pen not level.

(12) Split too wide ; trace too thick.

(13) Point too sharp, trace too thin ; pen digs into paper.

(14) Good medium trace.

(15) The point should be well burnished and slightly rounded to prevent the pen digging into the paper. No. 00 emery paper may be used for slightly rounding the point.

Fig. 6.20 Notes on recording pens from Negretti & Zambra's catalogue list 1928–30 (E5).

Department of Transport in Canada which gives an indication of some of these barograph pens. This circular reports that an abnormally large number of replacement instrument pens of a capillary tube type had been requested by stations for Meteorological Service of Canada barographs, and that only a small percentage of replaced pens had been returned to the headquarters. There had been problems with officers using the wrong instrument ink. These capillary pens employed Bristol instrument ink, said to be somewhat thinner than the Negretti & Zambra or other inks used in open-style instrument pens. The pen also had problems if it was allowed to dry out or if dust was allowed to contaminate the ink. As the capillary was so fine, it would clog up if not continually wet. There were also instructions on the size of the wire to be used to clean the capillary – some even used human hair, the inside diameter of the capillary being 5/000s of an inch – and they were made of platinum iridium tube. Of course, these days disposable fibre pens have replaced these. Although to my eye the throw-away plastic pens are not so attractive on a finely engineered instrument, I have to admit that they can be much simpler and less messy to use than the traditional style.

Although most of the traditional ink nibs are of the triangular bucket pattern seen in *Fig.* 6.21, occasionally one comes across a more substantial nib, which is shaped and rounded and sometimes called a boat nib (see *Fig.* 6.22). I fancy Negretti & Zambra may have made them as I tend to think I have noticed them more on their instruments that have come in for repair over the years. They all appear to be quite short so that the arm needs to be of the correct length to get the nib into the best position on the drum. Another type is perhaps best called a beak nib (see *Fig.* 6.23). There are two small triangles on the nib which touch the paper at 90 degrees, and ink is put into the open-ended part of the nib. I do not like to use these as I think they are somewhat messier than the normal type. After many years of making and repairing barographs, I rather think that Richard Frères had the best design and their style of nib looks better than any later type I have seen so far.

All types of nib appear to suffer from corrosion over the years as the ink is mildly corrosive, and tends to clog them up. There is also often corrosion of the ends of the aluminium arms due to a reaction between the ink and the metal. Later, stainless steel was used, and indeed our replacement nibs are made of stainless steel, although for authenticity of replacement arms we will often use aluminium as per the original.

Some years ago, while F. W. Darton & Co. were still trading from Bushey in Watford, I noticed that their barographs and recording instruments were supplied with an absorbent fibre that could be recharged with ink when needed. This absorbent fibre was held in a simple metal frame, as the normal nibs are, but without the triangular shape, just a very small round tube

Fig. 6.21 A traditional Richard Frères bucket type nib.

Fig. 6.22 A boat type nib.

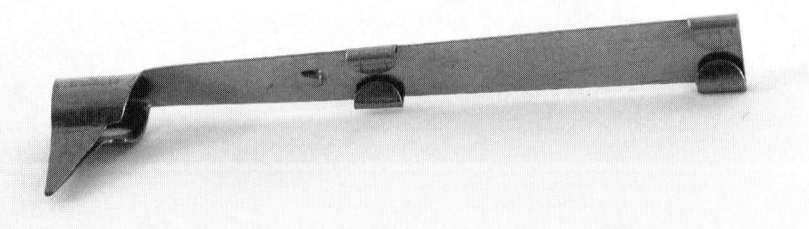

Fig. 6.23 A beak type nib.

Fig. 6.24 A modern disposable fibre nib.

into which a small piece of absorbent fibre, pointed at the paper end, was inserted and held firm. To recharge the fibre, you put a small drop of ink (yes, still messy if it dropped onto the barograph) onto the fibre which soaked it up and recharged it for another week or so, the advantage being that a more consistent line was traced. The disposable fibre nib that we are used to today (see *Fig.* 6.24) has, I believe, since replaced these. In our own workshops, we prefer to fit ink nibs on all old barographs that used them, as well as on our more traditional copies of the English barograph, although one always needs to have care with ink, absorbing any spills and wiping them up.

To fit a standard nib, you need first to ensure that the arm is good enough to be of use and not weakened by corrosion. Old arms of the aluminium type are tapered and vary in size at the end where the nib fits: there was no standard in previous times and no prescribed standard even today. The arm should be thin enough and tapered so a new nib can be fitted. With the thinner nickel silver or stainless steel arms, the new nib more often slides onto the end of the arm and then, using a pair of pliers, the two tabs are squeezed so as to clamp onto the arm; generally, a good tug will remove it again, if needed. On the aluminium arms, a little more work is often required. If the arm is damaged or too short, then a new arm may be needed.

When set up correctly, the point of the nib should be on the part of the drum closest to the front of the glass, although a little less does not make much difference. The arm may need carefully filing thinner on the taper, and the edges can be slightly angled so as to slide more easily into the nib. Generally, when fitting a nib, I find that the two tabs need carefully opening up so the end of the arm slides into them. This is a fiddly little task: the use of fine tools and a small vice is a great help (*Fig.* 6.25). The jaws of the small clamp press the tabs together on the end of the arm, and then the end of the tabs can be carefully tapped with a small hammer or pressed with the end of a screwdriver to wrap around the arm, before a pair of pliers can be used to clamp the tabs firmly on the arm.

Charts

Following the development of the Richard Frères barographs, which used ink containing glycerine, charts made of special paper became necessary to receive this type of ink. Some barographs used pencil recording, which could be used on many different types of paper, but the use of an ink which did not dry very quickly had a wetting action on most papers. Therefore, ordinary paper is of little use and a special quality of paper is needed.

One of the best papers for barograph charts used to be produced by a paper mill in Germany, although it is now no longer available. I believe it had 2 per cent plastic fibres woven into it during manufacture and did not allow the ink to absorb so readily. The most suitable paper generally

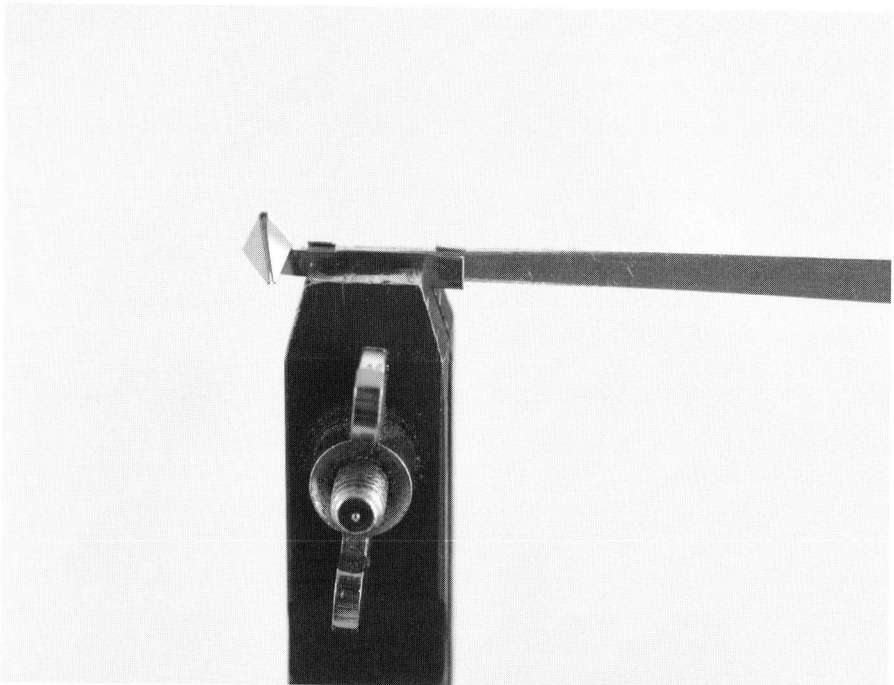

Fig. 6.25 Holding a nib to fit the arm into.

available since then comes from the USA and is what is known as a 'coated' paper. This makes reproducing charts more expensive than they would be if standard paper were used. A further problem with coated paper is that sometimes the coating is not applied consistently and misses a small section, leading to areas on a chart where the ink bleeds. I was informed by the suppliers of this coated paper that there is no way of telling if the paper is coated correctly after it has been made.

Many barograph users have tried photocopying charts when running short or if they consider the cost of specially printed charts too dear, but this usually causes the ink to run out in the same way as it would on blotting paper. It has been suggested that a certain type of computer paper is a good substitute. Photographic quality may also work well, but is so expensive as to make the cost of charts even dearer than having them printed on good paper. The main problem arises when only a few sample charts are required and it is uneconomic to have just a few dozen printed.

The small, brass carriage clock-type barograph featured in chapter 4 (*Fig.* 4.11) created quite a problem in the workshop. The customer wanted charts for it (but only a few) and the right paper was very difficult to obtain unprinted. In those days, we managed to reproduce charts for this model by means of a computer, using the back of ordinary charts.

In recent years, I have learnt a lot about paper and tested many different types. I eventually discovered a specialist paper mill that would mill paper for us – the snag was that we had to order 5 tons! So we did. We now have a lot of paper and hope to offer good-quality barograph paper for many years to come. By having paper specially milled for us that is less absorbent than other types of paper, we have a quality of paper that works well – provided that the ink is also suitable. No supplier can guarantee that their charts suit someone else's ink.

Most charts today can be reproduced if there is a good original, and we prefer to print with green or red ink in the same way as some of the old instruments were originally fitted up. The problem is still the print setting-up costs and it is seldom worth printing just a few hundred charts. Most charts will be copyrighted and may have some maker's mark on them so that copies can be identified. We always have the artwork drawn for us: this is expensive but gives us the chance to make old-looking charts for the models we restore or for customers' old instruments. Generally, it is not worth printing a fresh chart unless we have 50 packs printed.

If you do not have a correct chart for a rare barograph, then it is back to testing the time divisions needed on the chart by fixing a piece of paper on the drum, winding it and recording the distance travelled in seven days (if a seven-day clock, of course). Then one needs to pressure test the range of the arm – not all are the standard 3 inches. With these two measurements and the radius of the arm, a chart can be drawn that should suit.

The tendency is for modern charts to be made of white paper with the chart area printed in grey, which is fine for modern instrumentation, but old instruments have a charm which is spoilt by placing a modern-looking chart on them.

Originally, Richard Frères sold barographs that used a clip on the drum to hold the chart. In 1902, Short & Mason introduced a gummed chart which sits flatter on the drum and does away with that sometimes fiddly job of holding the paper around the drum while positioning the chart clip – easier when you have done it many times. The clip should not interfere with the nib as it rotates over it as the nib should be angled sufficiently to simply slide over the clip as it revolves around. The introduction of adhesive charts was quite a good idea, but does add cost over the lifetime of a barograph as adhesive charts are naturally more expensive.

Some years ago, it was possible to buy charts ready gummed, but then the company changed to double-sided tape and would charge £12 extra a pack for this, which was rather too dear for many people. They then started offering reels of double-sided tape for barograph users to cut and stick on their charts themselves – a time-consuming and tedious job. Ever interested in authenticity, I decided to investigate the gumming of charts, and since then we have gummed many thousands by hand. It is simple to do – just rather fiddly and tedious. The glue traditionally used is gum Arabic, the same glue used for many years on stamps. Once painted on the end of the chart and allowed to dry, the glue becomes sticky again when moistened. As the new paper we introduced was such a good paper for not absorbing ink, we had to alter the glue to be slightly more absorbent on the paper.

A little trick that can often assist in removing an adhesive chart at the end of the week is to fold over the ends with the perforations to weaken the paper just enough between the perforations to make the join (when stuck on the drum) easier to rip off.

If you are fortunate to have an early pencil-recording barograph, then almost any paper will be satisfactory to produce charts, and the nightmare of paper quality is not one that need worry you.

Ink

The basic ingredients of traditional barograph ink are water, glycerine, acetic acid (vinegar) and colouring. However, barograph ink needs to be matched to the paper used for the charts and the balance of ingredients is very fine to achieve the result required. I made many attempts at mixing before I eventually discovered what would work on our new paper as well as on most of the charts we had been selling before this. Testing ink and charts takes many months as one needs to test several variations on a barograph chart revolving at the normal speed.

Many years ago when I was searching for suppliers of charts and ink, I happened to speak with a soon-to-retire member of staff at a chart suppliers from Germany who informed me that he understood there was once a time when there were 90 different types of ink available. Nowadays, the choice in the industry for various recording instruments – if they still use this wet ink writing system (most are fibre pens or use electronic recording) – is slow, medium or fast-drying ink. We make a slow-drying ink for barographs using different ingredients from those used in modern ink. However, one can never guarantee that on odd occasions ink will not show some bleed on a paper as with original charts and ink: the very nature of what one is trying to achieve means it is bound to happen sometimes.

Appendix: Summary of Patents

Date	Number	Outline
1868	61	Self-recording Bourdon tube connected to pencil with clock drum
1868	2924	Chart recording thermometer and barometer using pencil
1869	2587	Pressure recording with pencil on paper drum
1871	1680	Marking trace by a dot through carbonised paper
1871	2596	Self-sharpening pencil recorder (part 4 of patent)
1874	1587	Magnified scale using moving apparatus aneroid or mercury tube
1877	1639	Recording pressure using electric current
1880	139070	Improvements in barographs: holding paper on drum and nibs to hold ink (French)
1885	9028	Electromagnet used for recording and transmitting pressure readings
1891	7323	Wall-mounted barograph with sprung chart, frictionless recorder (French)
1894	1900	Disc recording charts
1894	10791	Continuous paper roll with sprocket holes (steam pressure recorder) (USA)
1897	24417	Use chart seven times by moving pen position
1900	733	Light-sensitive paper recording

Year	Number	Description
1902	3715	Raised clock and glued charts
1902	12476	Electric recording apparatus, aneroid drives slave motor by electrical contacts
1902	27683	Temporary recording using a thread between two bands, one being clear
1903	11548	Rigid ended capsules
1904	22556	Use of aneroid underneath slab and easier to adjust (Short & Mason)
1904	428606	Registered (not patented) for dial on barograph (Short & Mason), 11 March
1905	18320	Endless steel chain held on by magnets on disc or strip chart (German)
1911	7500	Easy adjustment of pen for large variations in height
1911	20800	Mercury tube recorder with rod floating in vacuum
1916	108775	U-shaped arm to transmit record from movement to chart, enabling large chart with movement underneath
1919	145882	Straight line with cranked arm
1922	198838	Improvements in vacuum chambers: change in capsule soldered edge
1927	307114	Photographic type of recording level of liquid on chart
1936	501337	Balanced elasticity capsule for self-compensating for temperature
1949	641127	Oil dampening for use on ship

Bibliography

Air Ministry Meteorological Office (1936) *A Fishery Barograph: A Note on the Use of the Barograph in Anticipating Gales and Instructions for Care and Maintenance of Barographs Lent to Fishing Communities.* Meteorological Committee (MO 333).

Banfield, Edwin (1991) *Barometer Makers and Retailers 1660–1900.* Trowbridge: Baros Books.

Banfield, Edwin (1993) *The Italian Influence on English Barometers from 1780.* Trowbridge: Baros Books.

Banfield, Edwin (1996) *Barometers: Aneroid and Barographs.* Trowbridge: Baros Books.

Banfield, Edwin (2008) *Antique Barometers: An Illustrated Survey*, second edition. Trowbridge: Baros Books.

Belville, J. H. (1858) *Manual of the Mercurial and Aneroid Barometers.* London.

Bibby, J. R. (1949) 'The use of barographs in ships', *The Marine Observer*, vol. 19, no.143, January, p. 56.

Bolle, Bert (1981) *Barometers.* Woodbridge: Antique Collectors' Club.

Brenni, Paolo (1996a) 'Nineteenth century French scientific instrument makers: Louis Clement François Breguet and Antoine Louis Breguet', *Bulletin of the Scientific Instrument Society*, no. 50, September, p. 21.

Brenni, Paolo (1996b) 'Nineteenth century French scientific instrument makers: The Richard Family', *Bulletin of the Scientific Instrument Society*, no. 48, March, p. 11.

Collins, Philip R. (1998) *Aneroid Barometers and their Restoration.* Trowbridge: Baros Books.

Collins, Philip R. (2016) *Care and Restoration of Barometers*, second edition. North Curry: Baros Books.

Cumming, Alexander (1766) *The Elements of Clock and Watchwork, Adapted to Practice.* London.

Fellow of the Meteorological Society (1849) *The Aneroid Barometer: How to Buy and Use It.* London.

Insley, J. (2000) 'Instruments well adapted to the work: meteorological instruments in 1850 and since', *Weather*, vol. 55, no. 8, August, p. 254.

Jagger, Cedric (1983) *Royal Clocks.* London: Robert Hale.

McConnell, Anita (1988) *Barometers.* Aylesbury: Shire Publications.

McConnell, Anita (2013) 'Review of the exhibition *Dal Cielo alla Terra,*

Florence, 2013', in *Royal Meteorological Society History Group Newsletter*, issue no. 1.

Maxant, B. J. (2000) *L'Histoire du baromètre*.

Middleton, W. E. Knowles (1968) *The History of the Barometer*. Trowbridge: Baros Books, reprinted 1994.

Negretti & Zambra (1864) *A Treatise on Meteorological Instruments*. Trowbridge: Baros Books, reprinted 1995.

Ronalds, F. (1847) Paper in *The Royal Society's Philosophical Transactions*, vol. 1, pp. 111–17.

Thoday, A. G. (1978) *Barometers*. London: Science Museum.

Index

A&NCS Ltd, *see* Army and Navy Co.
Abraham, Adolf, 122
accuracy, 73, 89, 134, 138–40
Agolini, Giuseppe, 106–9
air transit, 134–5
altimeter, *see* altitude recorder
altitude adjustment, 31, 44, 45, 85, 88, 89, 135, 136
altitude recorder, 115–21
arm, *see* pen arm
Army & Navy Co./Stores (A&NCS Ltd), 51, 85, 125, 127
Artaria (Geneva), 123
Atmospheric Recorder (Dollond), 15, 16

balance weight, *see* counter-balance
Banfield, Edwin, 42, 116, 121–2
Barigo (Germany), 92, 95
Barker, John, & Co. Ltd, 46, 50
Barometer Makers and Retailers (Banfield), 42
Barometers: Aneroid and Barographs (Banfield), 116
Baromètre Holosterique, 124
Bartlett, Joseph, 21
base plate, 101–2
Bauerle, Tobias, and Sons, 87, 129
Beck, R. & J., 24, 25
beeswax, 141
bellows: accuracy, 73, 138–40; calibration, 138–40, 153, 156; cleaning, 143; concertina, 63, 64, 74, 75, 93, 110; horizontal, 126, 128, 129, 130; Negretti & Zambra, 79, 102, 145–7; repair, 137, 138; replacement, 51, 138; Richard Frères, 147, 148; *see also* capsules
Bibby, J. R., 68
blueing, 101, 141, 143–4
Bourdon, Eugene, 19, 26, 71, 98
brass cleaning, 134, 141, 143–4
Breguet, 15, 17, 156
Brontometer, 115
Brown, Edward, 15
Buckingham Palace, 1, 3, 4

calibration, 89, 90, 138–40, 147, 153, 156
capsules: double/twin, 46, 50, 71, 82, 92, 96, 122; number, 71, 73; single, 82; *see also* bellows
carriage, 34, 134–5
Casartelli, J., 6, 7, 8, 9, 40, 42
case work, 140; Richard Frères, 29, 106, 123–4
Casella, Louis P., 27, 28, 31, 79, 92, 93; catalogues, 9, 10, 21, 27–8, 115, 116
catalogues: Casella, 9, 10, 21, 27–8, 115, 116; Davis, 40, 41; Hicks, 20, 21; Negretti & Zambra, (early) 6, 18–19, 20, 21, 22, 26, 27, 28; (list E5) 145–7, 159, 160; (list M4) 153, 154, 155, 156
chart drawer, 44, 136
charts, 106, 134, 137, 163–6; adhesive, 55, 137, 165–6; changing, 136, 137; circular disc, 5, 96, 97, 121, 122; clips, 55, 58,

61, 64, 69, 76, 137, 165; continuous, 121, 122, 123; cyclo-stormograph, 85; design of, 112, 165; Kew barograph, 12, 13, 156; tall, 71, 72; vellum, 3, 4, 156; *see also* paper
clock, 15, 104, 122–3, 129; accuracy, 75, 134; care of, 133–4; Met. Office pattern, 63, 64, 69, 76, 77, 79, 81; raised (1902 patent), 55, 56; unusual, 42, 43, 121, 126, 128; *see also* time-marker
clock/barograph, 1, 4, 5, 6, 8, 9, 17, 18, 21–4, 105, 106, 107, 108
Collins, Philip (Merton), 31, 32, 61, 95, 150, 154, 156
concertina bellows, 63, 64, 74, 75, 93, 110
connecting rods, 153, 157, 158
Cooke, T., & Sons Ltd, 116
counter-balance, 44, 67, 150, 152, 154
C-spring, 51, 122
Cumming, Alexander, x, 1; barographs: 1–5, 156; *Elements of Clock and Watchwork*, 3, 5
cyclo-stormograph, 79, 82–5

Darton, F. W., & Co., 63, 92, 110, 161
Davis, Gabriel, 37
Davis of Derby, 37, 40, 41
Dent, E. J., 19, 98, 102
dial barograph: closed, 40, 42, 110, 111; open, 4, 40, 42, 56, 58, 79, 82, 98–101, 102, 121, 122; semi-circular, 96, 97, 102, 104, 156
diaphragm, *see* bellows
Dines, L. H. G., 13
Dines, William Henry, 13; barograph, 13–15
Dollond, G., 15; Atmospheric Recorder, 16
Dollond of London, 23

Elements of Clock and Watchwork (Cumming), 3, 5
Elliot Bros, 21, 37, 38, 94
Engineering Instruments and Industrial (Negretti & Zambra), 145–7
Extra Sensitive Barometer, 115, 116

fisheries barograph, *see* Meteorological Office
FitzRoy, Admiral, 79
flag indicator, 110
Fox Talbot, H., 11
French barographs, 19, 37, 85, 98, 104, 105, 106, 117, 123, 126, 130; *see also* Rédier, Antoine; Richard Frères
friction, 3, 15, 110, 126, 139, 145; jewelled movement, 98, 147; pen arm, 87, 104, 135, 138, 156

'gate'-type pen arm, 61, 63, 79, 82, 136
George III, 1, 2
Germany: altitude recorder, 118, 120, 121; barograph, 85, 87, 92, 95
Gluck Engineering, 31, 32, 92, 150, 151
Goerz, C. P., 122, 124
Great Exhibition, 15
Griffin & Tatlock Ltd, 71

heat damage, 135
Hicks, J., 104
Hicks, James J., 20, 21
hinged 'gate'-type pen arm, 61, 63, 79, 82, 136
History of the Barometer (Middleton), 1, 159
Hooke, Robert, 1, 107, 159

Horseman Gear Co. Ltd, 79, 81
Howard, Luke, 3, 4

ink, 136, 137, 166; colour, 68, 136
ink bottle, 54, 102, 117, 135–6, 147, 149

jewelled movement, 98, 134, 138, 147
Jordan, T. B., 11

Kelvin Bottomley & Baird Ltd, 56
Kew station barograph, 11–13, 156
keys, 30, 31, 85, 115, 118
King, Alfred, 9; barograph, 10, 21
Kuhn, F., 112

lacquering, 102, 134, 140, 141–3
L'Histoire du Baromètre (Maxant), 98
linkages, 137–8, 139, 143, 144
Lowther, James, 5
Lucking, James, 34–6
Lund & Blockley, 129

McConnell, Anita, 106
marine use, 28, 68, 76, 87, 121
Maxant, Bernard, J., 98, 106
mercury barographs, 1–15, 106–9, 129, 131–2
Merton barograph, 31, 32, 61, 95, 150, 154, 156
metal cleaning, 140–1
Meteorological Office: fisheries barograph, 31, 36, 61, 63, 64, 74–81; microbarograph, 64–9; pattern clock, 63, 64, 69, 76, 77, 79, 81; symbol, 76
Meyer, George, 21, 104
microbarograph, 64–9, 92
Middleton, W. E. K., *History of the Barometer*, 1, 159
Milne, Admiral Sir Alexander, 7, 8
Ministry of Defence, 68

Mottershead & Co., 15, 17
Mount Stewart House, 5
Munsey & Co. Ltd, 79, 82

Naudet, 123
Naylor, James, 123
Negretti, Henry, 26
Negretti, Paul Ernest, 26
Negretti & Zambra: bellows, 79, 102, 145–7; dial barographs, 98–100, 102, 104, 156; early barographs, 6, 7, 37, 39, 89, 90; early catalogues, 6, 18–19, 20, 21, 22, 26, 27, 28; jewelled movement, 98, 134, 138, 147; list E5, 145–7, 159, 160; list M4, 153, 154, 155, 156; 1920s/1930s' barographs, 71, 72, 87, 88, 112, 114, 115; numbering system, 87, 89; pen nib, 159, 160, 161; precision barograph, 73, 155; 'Regency' barograph, 147–51; self-recording aneroid, 20, 22, 23, 25; trademark, 147, 149
nib, *see* pen nib
Nichol, J. P., 11
Noble, Jack, 26, 145

oil, use of, 133, 137–8
oil dampening, 68, 144

paper, 163–6; photographic, 11, 13, 164; smoked, 15
Paris International Exhibition, 15, 26
pen arm, 135; fitting nib, 144, 163, 164; friction, 87, 104, 135, 138, 156; hinged 'gate' type, 61, 63, 79, 82, 136; replacement, 105, 135, 144; tension type, 56, 58, 61, 87, 135, 136; throw-off device, 30, 31, 34, 36, 44, 64, 66, 92, 94, 95, 133; unusual, 44, 46

pen nib, 112, 115, 135, 136, 159; fitting, 144, 163, 164; types, 159–63
pencil recording, 3, 4, 5, 8, 9, 15, 18, 163, 166
PHNB (Pertuis Hulot Naudet Berthod), 123
Pillischer, Jacob, 42, 43
Pillischer, Moritz, 42
Plath, C., 126, 128, 129
polishing, 140, 141
precision barograph, 73, 155
pressure chamber, 138–9, 144, 156

Rédier, Antoine, 21, 104, 106, 108, 129, 131–2
Richard, Felix, 26
Richard, Jules Nicolas, 26, 109, 110, 159
Richard, Max-Felix, 26
Richard Frères, 19, 20, 28, 51, 98, 153; altitude adjustment, 44, 135; arm throw-off device, 92, 94; bellows, 147, 148; case work, 29, 106, 123–4; chart clips, 55, 165; counter-balance, 150, 152; early barographs, 26, 27, 28–31, 33, 34, 35, 36, 37, 38, 115, 116, 140, 163; nibs, 159, 161, 162; 1920s' barographs, 85, 86, 109–10; Société Richard Frères, 26; trademark, 26, 33
Ronalds, Francis, 28
Royal Meteorological Society Newsletter, 106–9

Schultz, Emil, 121
Science Museum, 3, 4, 5, 9
self-recording aneroids, 15–25, 27
sensitivity of barograph, 73, 138–40
SGDG, 34, 116
Short, Frank, 145

Short & Mason, 28, 44, 55, 96, 100, 134; altitude recorder, 117, 119; barographs, 56, 58–61, 63, 64, 71, 73, 75, 102; charts, 55, 137, 165; cyclo-stormograph, 79, 82–5; 'gate'-type pen arm, 61; microbarograph, 64–9, 92; sales leaflet, 28, 29, 61; stormoguide, 110, 111; trademark, 40, 64, 67
siphon tube, 1, 2 , 8, 9, 13, 15, 129, 131
Société Richard Frères, 26
Statoscope, 115, 116
steel cleaning, 140
Steward, J. H., 51, 52
stormoguide, 110, 111
Switzerland, 112
Symons, G. J., 115

tension type pen arm, 56, 58, 61, 87, 135, 136
thermobarograph, 68, 70, 71, 72
throw-off device, 30, 31, 34, 36, 44, 64, 66, 92, 94, 95, 133
time-marker, 66, 74–5
Trotter, John, 100, 101

vellum chart, 3, 4, 156
vertical barograph, 105, 107, 124, 125
Victoria and Albert Museum, 5
Vidi, Lucien, 7, 19, 106

water gauge, 37, 40
'weather clock'/'weatherwiser', 1, 159
Weston, Edward, 129
Wilson Warden & Co., 28, 55, 56, 76, 153
Winter, T. B., & Son, 46, 47–9
Wren, Christopher, 1, 107

Yates & Son, 44, 47

Antique Barometers: An Illustrated Survey, second edition

Edwin Banfield

The author traces the development of the English domestic barometer from the earliest mercury stick instruments, through the elegant wheel barometers of the late eighteenth century to the thoroughly portable pocket aneroids of Victorian times. He also examines a variety of other instruments, including angle, double and multiple-tube barometers, marine, sea coast and Admiral FitzRoy barometers, aneroid barometers and barographs, with the aid of more than 130 illustrations.

Most types of antique barometer still available are illustrated, and useful information is given to assist in dating them. The book also contains advice on how barometers should be handled and maintained.

ISBN 978-0-948382-15-4
234 x 156 mm, x + 149 pp, 137 illus.
£9.95 (paperback)
First edition published 1976
Reprinted 1977, 1978, 1980, 1981, 1983, 1989, 1996
Second edition published 2008

Aneroid Barometers and their Restoration

Philip R. Collins

The author charts the beginning and development of the popular aneroid barometer, with many illustrations of both common and rare examples. He examines their mechanisms and gives a detailed account of how to repair them, as well as describing some of the case repairs often needed. With the aid of examples, he sets out to help date aneroid barometers and shows what a diverse selection has been made and sold over the past 150 years.

'Philip Collins [demonstrates] the wide variety of style and design available in aneroids, as well as setting them in the context of their development both as accurate scientific instruments and as *objets d'art* ... His enthusiasm and extensive knowledge of barometers of all types are impressive.' *Edwin Banfield* (from the Foreword)

ISBN 978-0-948382-11-6
234 x 156 mm, x + 212 pp, 305 illus., incl. 16 colour
£18.95 (hardback)
First published 1998

Barometers: Aneroid and Barographs

Edwin Banfield

The author traces the history of aneroid barometers, barographs, storm glasses and weather houses, with the help of 130 illustrations. He covers the range of barometers and barographs currently available and gives advice on how to date them.

This is the third volume in the 'Barometers' trilogy by Edwin Banfield, following the companion books *Barometers: Stick or Cistern Tube* and *Barometers: Wheel or Banjo*.

ISBN 978-0-948382-02-4
234 x 156 mm, vii + 150 pp, 130 illus.
£11.95 (hardback)
First published 1985
Reprinted 1996

Care and Restoration of Barometers, second edition

Philip R. Collins

In this second edition of *Care and Restoration of Barometers*, Philip Collins draws on almost 40 years' experience of working on antique barometers to show how to restore old instruments and thus increase their attractiveness, usefulness and value. With the aid of some 175 illustrations, he describes in detail the steps involved in restoring a wheel barometer from first dismantling the instrument, through the cleaning and repair of parts, refitting the mercury tube, case repair and refinishing to final reassembly. Separate chapters deal with the special problems that can arise with stick or cistern tube barometers, aneroid barometers and barographs. Advice is also given on how to handle and maintain antique barometers, once restored.

ISBN: 978-0-948382-16-1
234 x 156 mm, viii + 152 pp, 175 illus.
£9.95 (paperback)
Second edition published 2016